Developments in Geotechnical Engineering

Series Editors

Braja M. Das, California State University, Henderson, CA, USA

Nagaratnam Sivakugan, James Cook University, Townsville, QLD, Australia

This series on Geotechnical Engineering has been initiated to provide an integrated platform for publishing in all areas of geotechnics and their applications to engineering problems. The series focuses on the topics that have garnered recent interest from a large research audience, including, but not limited to, saturated and unsaturated soils, expansive soils, geosynthetics, marine geotechnics, and soil improvement. This series includes monographs, textbooks, and references in the general area of Geotechnical Engineering. The books in this series will serve researchers, practitioners, and students alike.

Hemanta Hazarika · Juichi Nakazawa ·
Iwao Nakahara
Editors

Practices and Trends in Ground Improvement Techniques

Springer

Editors
Hemanta Hazarika
Kyushu University
Fukuoka, Japan

Juichi Nakazawa
Tokyo Engineering Consultants Co., Ltd.
Osaka, Japan

Iwao Nakahara
Japan Foundation Engineering Co., Ltd.
Tokyo, Japan

ISSN 2364-5156 ISSN 2364-5164 (electronic)
Developments in Geotechnical Engineering
ISBN 978-981-19-3321-9 ISBN 978-981-19-3322-6 (eBook)
https://doi.org/10.1007/978-981-19-3322-6

This Springer imprint is published by the registered company Springer Nature Singapore Pte Ltd.
The registered company address is: 152 Beach Road, #21-01/04 Gateway East, Singapore 189721, Singapore

Preface

Practices and Trends in Ground Improvement Techniques is a compilation of twelve case histories from many countries around the world (five from Japan, two from the USA, and one each from Bangladesh, Cambodia, India, Indonesia, and Thailand). This book is born out of conviction that the discipline of ground improvement is mostly practice rather than theory. Although some ground improvements can be performed using the theoretical concepts of conventional soil mechanics, not all are so. Sometimes, strategic ideas are required to execute the techniques in the fields first, and then, the theoretical basis of those can be established.

Soil, as a natural material, is highly variable and is affected by numerous factors, and therefore, finding a unique and comprehensive solution for every problem the engineers encountered is a herculean task. This book is the results of authors' persistent desire to organize and present some selected case histories of ground improvement. The authors feel that case histories are extremely essential for the young researchers and engineers to develop an interest in the subject and contribute toward further advancement of this field through continuous engagement. This book is an effort in this direction. Numerous case studies, described in this book, show how the new and improved methods have actually been applied over a wide variety of ground conditions. It also focuses on the new applications and trends of the ground improvement geo-system including recycling, geo-environmental consideration, and preservation worlds' cultural heritage.

As engineers seek economical solutions dealing with myriads of problematic soils, the need to improve the poor-quality ground in urban sites increases, and those demands are expected to remain steady or increase in the future as more and more countries develop their infrastructures. Designers need to make judicious decision based on his/her experiences and facts available. In this book, ground improvement methods adopted in a particular situation and accomplishment of the projects are described in an easily understandable way. This book helps researchers and engineers to understand, how the principles of ground improvement methods are executed in the site, basis of selection of a particular ground improvement method in a project, cost-performance of such methods, etc. The mechanism, design, and execution procedures of various ground improvement techniques, which are described as case studies in this book, can be found in *"Essentials of Ground Improvement Techniques"* by the same authors.

This book is intended for the beginners and the intermediate-level practitioners who are dealing with geotechnical construction projects or who have interest in the development and practical application of ground improvement techniques. The book will be an excellent resource for those involved with design and execution of ground improvement projects. The authors believe that engineers and researchers will find it helpful in developing, advancing, and applying their techniques in the field.

Fukuoka, Japan — Hemanta Hazarika
Osaka, Japan — Juichi Nakazawa
Tokyo, Japan — Iwao Nakahara

Acknowledgments

The publication of *Practices and Trends in Ground Improvement Techniques* has been possible through the support and cooperation of many of the authors' colleagues and former students. The authors express their sincere thanks to the following individuals for their valuable time and effort in providing data, making suggestions for improvement, drafting figures, and translating documents from Japanese to English.

Prof. Hajime Imanishi, Prof. Emeritus, Tohoku Institute of Technology, Sendai, Japan.
Prof. Tadashi Hara, Kochi University, Kochi, Japan.
Mr. Nozomu Kosaka, former President, JAFAC USA, California, USA.
Mr. Toshiro Hara, formerly at Japan Foundation Engineering Co., Ltd., Tokyo, Japan.
Mr. Kengo Nishimura, Chugoku Electric Power Corporation, Hiroshima, Japan.
Dr. Siavash Manafi Khajeh Pasha, IMAGEi Consultant Corporation, Tokyo, Japan.
Mr. Katsuya Ogo, Nippon Koei Co. Ltd., Osaka, Japan.
Dr. Divyesh M. Rohit, Formerly at Kyushu University, Fukuoka, Japan.

Acknowledgements

Contents

About the Editors

Hemanta Hazarika is currently a Professor at the Graduate School of Engineering and Department of Interdisciplinary Science and Innovation, Kyushu University, Fukuoka, Japan. His research activities include disaster prevention and mitigation, stability of soil-structures during earthquakes and tsunami, ground improvement, application of recycled waste and lightweight materials in constructions, and landslides mitigations. He has more than 350 technical papers in various international journals, international conferences, workshops and symposia to date. He has also authored two textbooks on Soil Mechanics and their Japanese versions. In addition, he also served as the editor of three technical books on disaster mitigation and waste tire recycling. Prof. Hazarika pioneered the use of waste tire derived recycled materials for earthquake protection of infrastructures and holds two patents on those. Prof. Hazarika has several years of experience in teaching, research as well as geotechnical practice and consulting both within and outside Japan. He has also successfully led a World Bank project on "Improving the Resilience of Nepal's Strategic Roads Network". Currently he is also the technical advisor of Japan Foundation Engineering Co., Ltd., Tokyo, Japan, and President of GLOSS (Global Society for Smart Geo-Sustainnovation).

Juichi Nakazawa is a currently the technical adviser in Tokyo Engineering Consultants Co. Ltd. Tokyo, Japan. He obtained his Doctor of Engineering degree from Kyoto University in 1982. After graduation, he served as a research associate at Kyoto University for 5 years. He has a wealth of experiences as a geotechnical engineer for about 60 years, including his 30 years of engagement as geotechnical engineers in general construction company, and as researcher in National College of Technology, Maizuru, Japan where he worked as professor until his retirement. He was a member of the International Project Committee (UNESCO), which proposes the preservation method of historical heritages composed of soil and rocks. His research activities include soft ground treatments, stability of soil and rock slopes, foundations by cast in place pile, tunneling for problematic ground and preservation of cultural heritages. He proposed many solutions for ground with geotechnical problems, and those results were presented in more than 150 technical papers that he published. In addition, he served as a part-time lecturer at four universities of Japan, and at the international training center (JICA) with the aim of transmitting his experience to young engineers. He is a licensed Professional Engineer in Japan.

Iwao Nakahara is presently the president of Japan Foundation Engineering Co., Ltd., Tokyo, Japan. He obtained his master's degree in marine engineering from Tokai University, Japan in 1976. After his graduation, he worked as technical officer in the same university, during which he was involved in the research related to the mining of marine resources. Since his joining in Japan Foundation Engineering Co., Ltd., Mr. Nakahara has been involved in the development and operationalization of techniques for infrastructure development, ground anchoring, and tunneling reinforcement, etc. For the last fourteen years he has been serving as the president of Japan Anchor Association and has been actively contributing towards standardization and spreading of ground anchoring techniques and their maintenances.

Chapter 1
Introduction

Hemanta Hazarika

1.1 Prologue

Ground improvement is one of the basic requirements of many civil engineering construction projects. To meet the demands for improving previously unusable sites, researchers, practicing engineers and specialty contractors have been developing varieties of ground improvement methods and operating systems. As engineers seek economical solutions dealing with myriads of problematic soils, the need to improve the poor-quality ground in urban sites increases and those demands are expected to remain steady or increase in the future as more and more countries develop their infrastructures. Therefore, the field of ground improvement relies heavily on case studies, where past experiences of both success or failure are discussed, with emphasis on specialized research.

The discipline of ground improvement is mostly practice rather than theory. Although some ground improvements can be performed using the theoretical concepts of conventional soil mechanics, not all are so. Sometimes strategic ideas are required to execute the techniques in the fields, and later the theoretical basis of those can be established. Therefore, the need for case histories of ground improvement is extremely essential for the young researchers and engineers to develop interest in the subject and contribute toward further advancement of this field through continuous engagement. This book is an effort in this direction.

This book covers 12 case studies encompassing each of the following categories of ground improvement described in [1] as well as in the book "Essentials of Ground Improvement Techniques" by Hazarika et al. [2].

I. Ground Improvement without Admixtures in Non-cohesive soils or Fill Materials

H. Hazarika (✉)
Kyushu University, Fukuoka, Japan
e-mail: hazarika@civil.kyushu-u.ac.jp

H. Hazarika et al. (eds.), *Practices and Trends in Ground Improvement Techniques*, Developments in Geotechnical Engineering,
https://doi.org/10.1007/978-981-19-3322-6_1

(Compaction, Vibro-compaction and Vibro-flotation, Heavy Tamping)

II. Ground Improvement without Admixture in Cohesive Soils
(Replacement, Preloading using Fill and Vacuum, Vertical Drain)

III. Ground Improvement with Admixture or Inclusion
(Surface Compaction using Cement and Lime, Sand Compaction Pile, Stone Colum)

IV. Ground Improvement with Grouting Type Admixture
(Particulate Grouting, Chemical Grouting, Jet Grouting, Mechanical Mixing, Cement Deep Mixing)

V. Ground Improvement by Reinforcement of Soils
(Geosynthetics, Nailing, Anchoring)

VI. Other Methods of Ground Improvement.
(Electro-osmosis, Freezing, Low-carbon Techniques).

For details of those techniques, such as the theoretical principles and mechanism, procedures for implementing those, design methods, and applicable range of those, the readers are referred to that book. Case studies from Bangladesh, Cambodia, India, Indonesia, Japan, Thailand, and the USA. are included in this book, representing diverse geologies, infrastructural needs, economies, and technologies. Based on the authors' experiences of working in various ground improvement projects and their implementations throughout the world, this book also focuses on the new applications and trends of ground improvement techniques including recycling, geo-environmental considerations, and cultural heritage preservation.

1.2 Overview of the Case Studies

The first case study covers the category II: Ground Improvement without Admixture in Cohesive Soils. It uses Prefabricated Vertical Drains (PVD) with preloading in a Port Park area in the city of Chittagong, Bangladesh. The case study involves a container yard, which was constructed for handling of loaded containers at Port Park area of Chittagong sea port covering an area of 60,700 m^2 (15 acres). The sub-soil within the yard contained layers of very soft to soft strata of clayey silt/silty clay at depths of 1.0–3.5 m below the existing ground level. Improvement of the soft soils was a major challenge of this project.

The second case study covers the category III: Ground Improvement with Admixture or Inclusion. It describes the use of soil–cement method for improving jobsite road foundation. This particular case study involves the construction project of Tokai-loop expressway, which connected many cities in and around Nagoya, Japan. Many jobsite roads were needed for carrying out the operation of construction machines. Surplus soils from jobsites were also needed to be used effectively. The trafficability of the soft ground of those temporary roads was improved by adopting soil–cement method.

The third case study covers the category III: Ground Improvement with Admixture or Inclusion. It describes two examples of application of Sand Compaction Pile (SCP) method in the United States. Those were aimed toward liquefaction prevention, bearing capacity improvement, and prevention of lateral spreading of ground. Postconstruction field testing showed sufficient improvement effects in both the projects.

The fourth case study covers the category IV: Ground Improvement with Grouting Type Admixture. It describes ground improvement using Cement Deep Mixing (CDM) method, which was designed to reinforce the soft ground for the emergency repair of a levee along the west bank of the Mississippi River in the United States. The long-term settlement of the remediated levee, which may occur due to the compression of the soft foundation soils, could be controlled using this method.

The fifth case study also covers the category IV: Ground Improvement with Grouting Type Admixture. It describes jet grouting method, which was used to construct clay-cement piles to improve the soft foundation made of Bangkok clay. This case study examines the behavior of the hexagonal wire mesh reinforced embankment constructed on jet grouted cement mixing piles to improve soft ground. Surface settlements, subsurface settlements, lateral movement, excess pore water pressure, and tension in hexagonal wire mesh were evaluated through observed field data and numerical simulation.

The sixth to ninth case studies cover the category V: Ground Improvement by Reinforcement of Soils. The sixth case study elaborates the use of geosynthetics on the challenging geotechnical problems of building 15–37 m high slopes on clay shale formations. The effectiveness of geosynthetic reinforcement is described in detail through discussion on the strength degradation of clay shale, the design aspects, and construction process of the geosynthetic reinforced soil slopes.

The seventh case study the category V: Ground Improvement by Reinforcement of Soils. The seventh case study describes a challenging project of road widening by cutting slope and stabilizing the slope using ground nailing. Project could be accomplished without causing any damage to the existing infrastructures, such as railway tunnels and residential houses.

The eighth case study the category V: Ground Improvement by Reinforcement of Soils. The eighth case study is an example of environmentally friendly sustainable ground improving technique in which *ecofriendly and biodegradable* natural materials were used as geotextiles. Coir geotextiles, which were used for erosion control of embankments, could serve the intended design purpose.

The ninth case study covers the category V: Ground Improvement by Reinforcement of Soils. In the ninth case study, two case studies are described in which industrial wastes (waste tires) were used to protect embankments. Through the field trials, laboratory investigation, and numerical simulations emergence of such environmentally friendly sustainable materials is demonstrated.

The tenth case study covers the category VI: Other Methods of Ground Improvement. This particular ground improvement technique deals with SDGs (Sustainable Development Goals) of the United Nations. To mitigate global warming by storing carbon, the log piling method for liquefaction mitigation and carbon stock was applied in residential areas in Chiba, Japan, which suffered tremendous damage from liquefaction due to the 2011 Great East Japan Earthquake. The method could ultimately store approximately 790 tons of CO_2 while improving the ground against earthquake-induced disasters.

The eleventh case study covers the category VI: Other Methods of Ground Improvement. Here two case studies are described, in which ground freezing method was used as temporary protection works in subway tunnel and discharge tunnel in thermal power plant station. The salient features of this improvement technique are reducing environmental loads and shortened construction.

The twelfth case study covers the category VI: Other Methods of Ground Improvement. This case study deals with the preservation of historical monuments through combined use of slaked lime and geotextiles. And, thus, this method is a combination of category III and Category V. To increase the bearing capacity of Angkur ruins foundations due to fluctuating water table, the original sandy soils were mixed with slaked lime and compacted by tamping with geotextile. The factor of safety of the bearing capacity could be improved to 1.5 by adopting this technique.

References

1. Hazarika, H., Nakazawa, J.: Classification of modern ground improvement techniques—A mechanism based approach. In: Proceedings of the 13th Symposium on Ground Improvement, The Society of Material Science, Japan, pp. 235–240 (2018)
2. Hazarika, H., Nakazawa, J., Nakahara, I.: Essentials of Ground Improvement Techniques. Springer Nature (To be published) (2023)

Chapter 2
Ground Improvement Using Prefabricated Vertical Drains with Preloading for Port Park Area at Chittagong, Bangladesh

Abu Siddique, Ashutosh Sutra Dhar, and Syed Fakhrul Ameen

2.1 Introduction

A container yard was constructed in 2009 at Chittagong Port, the largest sea port in Bangladesh, for handling of loaded containers. The site is located on the bank of the Karnafully river beside the Bay of Bengal in the Indian Ocean. Figure 2.1 shows the location of the site along with the surrounding geological and geomorphologic features. The yard covered an area of 60,700 m^2 (15 acres) and was designed to support a container load producing a contact pressure of approximately 56 kPa. The site is locally known as Port Park. Comprehensive field and laboratory investigations were conducted on sub-soils at the Port Park area to evaluate relevant geotechnical design parameters for the design of ground improvement. Field investigations revealed the presence of a very soft to soft clayey silt/ silty clay layer at the depths of 1.0–3.5 m below the existing ground level (EGL). The thicknesses of the soft layer varied from 3.0 to 7.0 m. A ground improvement scheme was designed and implemented to pre-consolidate the soft sub-soil before construction of the container yard to keep the settlements of the container yard within tolerable limits during the service life. Earlier [7] reported sub-soil conditions and details of design of the ground improvement works using preloading with prefabricated vertical drains. This chapter presents a summary of the sub-soil conditions at the project site, the geotechnical aspects for the design of ground improvement, ground improvement methodology,

A. Siddique (✉) · S. F. Ameen
Bangladesh University of Engineering and Technology, Dhaka, Bangladesh
e-mail: abusid@ce.buet.ac.bd

S. F. Ameen
e-mail: fameen@ce.buet.ac.bd

A. S. Dhar
Memorial University of Newfoundland, St. John's, Newfoundland and Labrador, Canada
e-mail: asdhar@mun.ca

H. Hazarika et al. (eds.), *Practices and Trends in Ground Improvement Techniques*, Developments in Geotechnical Engineering,
https://doi.org/10.1007/978-981-19-3322-6_2

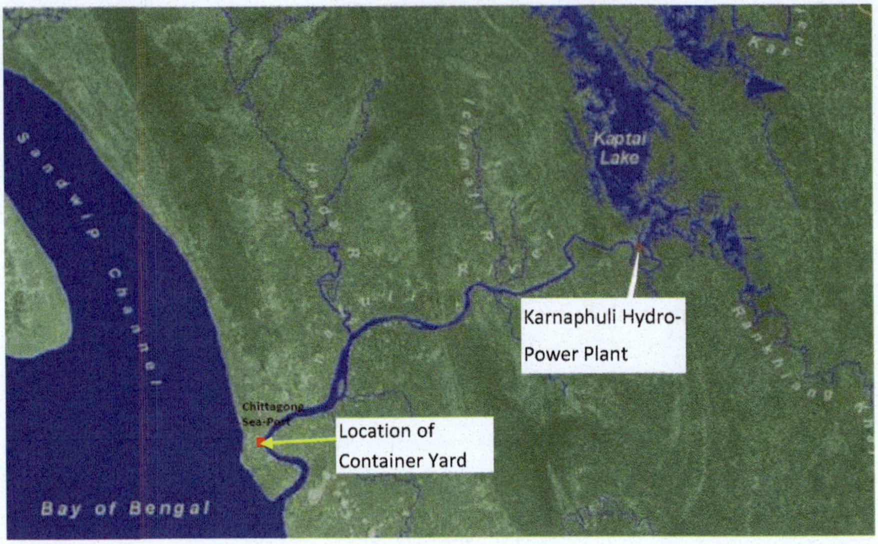

Fig. 2.1 Location of the site along with adjoining geological and geomorphologic features (after [7])

an evaluation of the ground improvement works through field monitoring, and findings from the field monitoring regarding consolidation with prefabricated vertical drains.

2.2 Sub-soil Conditions at the Project Site

The site for the container yard at Port Park area is a tidal plain as a narrow strip between Chittagong hilly uplands and the Bay of Bengal. The surface geology of the site is mainly governed by shallow sea water and the flood plain activities of the river Karnafully and its tributaries. The sub-soil includes very soft to firm silty clay or clayey silt and fine-grained silty sand with some decomposed materials near the ground surface.

The field investigations at the container yard in Port Park area consisted of drilling of boreholes, recording density/stiffness characteristics of sub-soil layers by carrying out Standard Penetration Test (SPT), collection of disturbed and undisturbed tube samples from different depths of the boreholes. A total of 15 boreholes, distributed over the area, were drilled vertically using wash boring technique. Wash borings of small diameter (approximately 125 mm) were drilled by water flush aided by chiseling. The approximate locations of the boreholes are shown in Fig. 2.2. The boreholes were drilled up to depths ranging from 14 to 24.5 m below the existing ground level (EGL). The density and stiffness characteristics of the sub-soil layers in the boreholes were measured by performing Standard Penetration Test (SPT) at

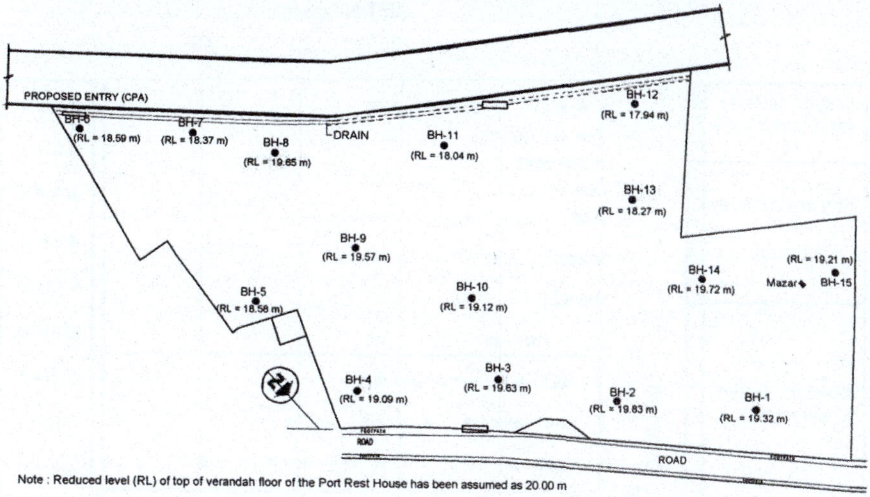

Fig. 2.2 Approximate locations of boreholes drilled in Port Park area (after [7])

1 m intervals up to depth of 5 m and at 1.5 m intervals thereafter by means of standard 50.8 mm outside diameter split-spoon sampler. Undisturbed samples were also retrieved from cohesive layers by pushing conventional 76 mm external diameter thin-walled shelby tubes.

The method of the geotechnical investigation was chosen based on the technology locally available in Bangladesh. It deserves mentioning that Cone Penetration Test (CPT) facilities with the piezocone probe are not readily available in Bangladesh. In addition, shelby tube samplers were used to collect undisturbed soil samples for this project since Osterberg or other piston samplers were not available to the drilling contractor. However, all shelby tube and split-spoon samples were examined visually to identify the presence of any localized features (e.g., sand seams) and the consistency of the cohesive soil. An extensive laboratory investigation was then carried out for characterization of sub-soils and for determination of geotechnical design parameters.

Figure 2.3 shows generalized sub-soil conditions obtained from the geotechnical investigations. Ground conditions at the site were found to vary widely. Generally, the soil at the ground surface was fill materials consisting of light brown clayey silt or brown silty sand/sandy silt. The clayey silt was in a state of firm to stiff consistency. The silty sand or sandy silt was in a medium dense state. The fill materials extended from the ground surface and continued down to the depths of 1.0–3.5 m below the EGL.

Layers of very soft to soft cohesive silty clay/clayey silt were encountered beneath the fill materials. The thickness of the very soft to soft clayey silt/silty clay varied from 3.0 to 7.0 m and extended down to the depths of 5.0–8.6 m below the EGL. Field SPT N-values encountered in the very soft to soft silty clay/clayey silt layers were found to vary between 0 and 3. These very soft-to-soft deposits were highly

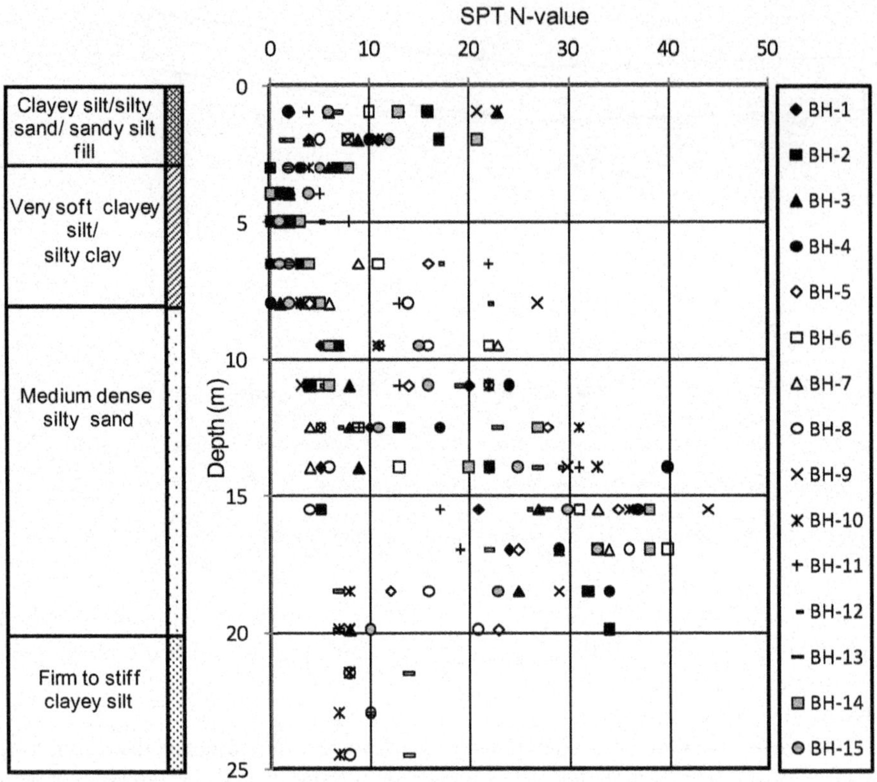

Fig. 2.3 General sub-soil profile along with SPT *N*-value (reproduced from [7])

compressible and expected to cause excessive settlements to the container yard under service loads. The silty clay/clayey silt layers were also encountered scattered at different depths in few boreholes. In general, the very soft to soft cohesive layer was underlain by a silty-sand deposit which was subsequently underlain by a layer of silt and/or silty sand with occasional laminations of clay. State of compactness of the silty-sand layers at the site was found to vary widely from loose to dense. The positions of groundwater table within the project site vary widely. The positions of groundwater table have been found to vary from 0.3 m (below the EGL) to 3.40 m (below the EGL).

Some of the findings and conclusions based on laboratory investigations are summarized below:

- The values of liquid limit of the samples obtained from the 12 boreholes varied between 32 and 56. Plasticity index varied between 9 and 25.
- The values of natural water content of the undisturbed tube samples were found to vary between 26.6 and 57.3%.

- Percentage finer (0.075 mm) of the nine samples ranged between 10.6 and 26.2%, while the fineness modulus ranged from 0.51 to 1.35.
- Undrained shear strength (undisturbed samples) ranged between 9 and 56 kPa with an average value of 26 kPa.
- Undrained shear strengths from Unconsolidated Undrained (UU) tests of the undisturbed samples reveal that for very soft samples they ranged between 7 and 18 kPa, while for soft samples they ranged between 21 and 30 kPa.
- From Consolidated undrained (CU) tests, the values of undrained cohesion (c_u) of the samples were found to vary between 10 and 35 kPa, while the values of undrained angle of internal friction (ϕ_u) were found to vary between 12° and 26°, respectively.
- Compressibility and permeability properties of 12 samples obtained from nine boreholes reveal the following.

Coefficient of volume change (m_v): 1.30×10^{-4}–49.5×10^{-4} m^2/kN.
Coefficient of consolidation (C_c): 0.17–0.45.
Initial void ratio (e_0): 1.04–1.62.
Based on the above values, the design values of c_v, c_h, k_v, and k_h were obtained, which are as following.

c_v: 7.5 m^2/year
c_h: 15.5 m^2/year
k_v: 1.5×10^{-9} m/s
k_h: 2.3×10^{-9} m/s.

2.3 Ground Improvement Scheme

The presence of very soft to soft clayey silt and silty clay deposits at the Port Park area indicates the strong possibility of substantial settlement unless effective measures for improvement of sub-soil are undertaken before the construction of pavement for the container yard in order to avoid future failures/problems. Alternatively, deep foundations can be used to transfer the load to dense sandy stratum. Of these above two alternatives, namely, ground improvement and construction of deep foundations, it appeared from preliminary estimation that deep foundations would be very costly compared with ground improvement. Therefore, improvement of sub-soils at the Port Park area of Chittagong Port should be adopted using a suitable technique. The selection of most suitable technique for any ground improvement project depends on four main considerations, namely, type of structure, type of soil, environmental considerations, and economic considerations. Methods and applications of ground improvement techniques in Bangladesh have been summarized by Ansary and Doulah [1]. Typical methods used to increase the bearing capacity of soft soils of Bangladesh include preloading and vertical drainage (sand drain and prefabricated vertical drain) in conjunction with preloading. Loading berms, or pressure berms, are being used as counterweights to increase the stability of road and railway embankments in

Bangladesh. Fabric-reinforced soil, where woven or non-woven fiber materials are placed in road and railway embankments to improve their overall stability. Sand columns or sand compaction piles have also been used in a number of instances for improving the underlying sub-soil. Several case studies regarding ground improvement for construction of buildings, roads, and embankments were reported by Reference [11] to illustrate the important features of ground improvement methods in soft clay deposits of Bangladesh. Due to lack of mechanical skills, equipment, and high costs, other ground improvement techniques like deep compaction (dynamic consolidation and dynamic replacement, blasting and vibro-flotation), admixture stabilization, stone columns, vacuum consolidation, deep mixing method, electro-osmosis, reinforced earth method and lime piles and columns are yet to gain popularity in Bangladesh.

When a layer of soft clay is very thick or the permeability is exceptionally low, the preloading technique is likely to be inefficient when used alone because an inordinately long period of time will be needed to bring about significant consolidation. Under these circumstances, radical reductions in the preloading time can be effected by installation of vertical drains to shorten the drainage path under which the clay consolidates. This has become a commonly employed device with soft clays in Bangladesh. The benefits of vertical drains used in conjunction with preloading are related almost entirely to acceleration of the primary consolidation of the clay since they bring about the rapid dissipation of excess pore pressures. This method is very effective for improvement of soft clays. Considering the nature of the sub-soils at the Port Park area of Chittagong Port, it is presumed that preloading with sand drain or prefabricated vertical drain (PVD) would be an effective measure for the improvement of existing soft sub-soils. Ground improvement works needed for the soft sub-soils at the Port Park area of Chittagong Port have been designed based on geological profiles from field tests and results of laboratory tests. Design of soil improvement works involved the following steps:

- Determination of maximum thickness of soft soil layer from the bore log data.
- Estimation of the magnitudes of consolidation settlement of soft soil layer.
- Estimation of time for consolidation settlement for achieving an acceptable degree of consolidation. If the consolidation settlement time to achieve 90% consolidation is reasonably short, then soil improvement can be done by preloading only. Otherwise, preloading with installation of sand drains or PVDs will be required to reduce consolidation time.
- Estimation of consolidation time and spacing of sand drains or PVDs to achieve anticipated settlement prior to pavement construction.

2.3.1 Consolidation Settlement and Time for Consolidation Settlement

Estimated maximum settlement was 536 mm for the 7 m thick layer located 2 m below the EGL while estimated maximum settlement was 203 mm for the 5 m thick layer located 3 m below the EGL. These settlements are too large for serviceability consideration of the container yard. It was therefore considered to pre-consolidate the soil before construction of the yard. Use of preloading for pre-consolidation was first considered due to the simplicity of the method and the suitability for implementation using local technology. However, the time required for the consolidation using preloading appeared as a major concern in the design of ground improvement. With the consolidation property estimated from the laboratory tests, (i.e., design value of $c_v = 7.5\ m^2/year$) and considering single drainage, the time required for 90% consolidation was estimated to be 1 year and 5.5 years for 3 m thick and 7 m thick layers of soft soil, respectively. Preloading with vertical drains (sand drains or PVD) was therefore considered for the design of the yard to accelerate the consolidation process. A brief description of the design concept and construction for ground improvement using vertical drains with preloading are presented in the following sections.

2.3.2 Design Assumptions for Ground Improvement

The effects of vertical drain on the consolidation are generally analyzed using an idealized model shown in Fig. 2.4. In this model, the vertical drain has been idealized as an equivalent circular drain. An annular zone, called a smear zone, is considered in the soil surrounding the drain to account for the disturbance caused by the installation of the drain. The permeability of the smear zone in the vicinity of the drain is reduced compared to the original base soil due to installation disturbance. Several methods are available to account for the smear effects in the design [8, 9, 12, 13].

In order to examine the smear effect in the design for ground improvement of the container yard, diameter of the smear zone was considered as two times the equivalent diameter of the drain (i.e., $d_s = 2d_w$) as per recommendation of [2].

2.3.3 Design of Vertical Drains

Soil improvement works for the Port Park area of Chittagong Port were designed based on the information on geotechnical profiles and the results from laboratory tests. Maximum thickness of the soft soil layer (i.e., 7 m) was conservatively considered for the design. Vertical drains were designed to install down to the depth of approximately 9 m below the ground level to cover the full depth of the soft clay layer.

Fig. 2.4 Idealization of consolidation using vertical drain (after [6])

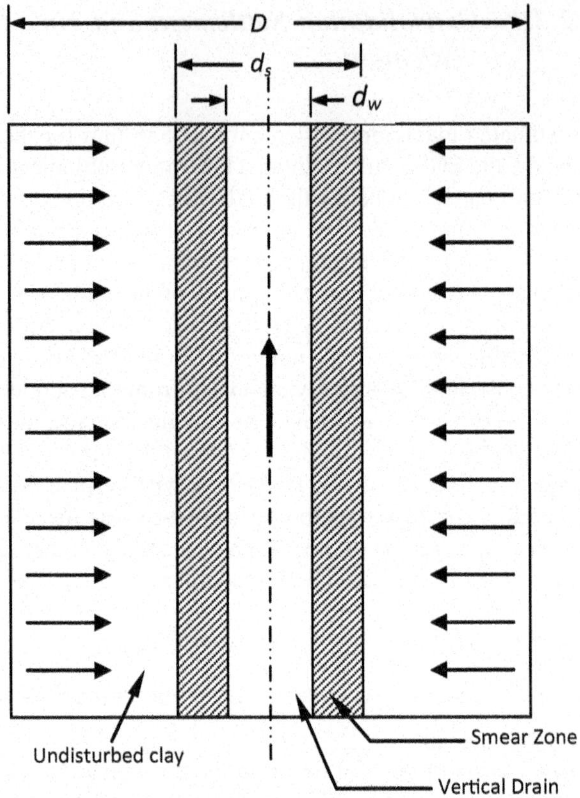

Table 2.1 shows three different options of vertical drains initially assessed. Time for 90% consolidation with 200 mm diameter sand drains at 1.5 m center-to-center spacing in square pattern was estimated to be 80 days, which is significantly less than the estimated period of consolidation without vertical drains (1 year and 5.5 years

Table 2.1 Approximate cost and time for three options of soil improvement (after [7])

Option	Consolidation time (days)	Approximate cost of vertical drain (Taka/sft)
(1) 200 mm diameter sand drain @ 1.5 m c/c in a square pattern	80	50
(2) PVDs (width = 100 mm, t = 4 mm) @ 1.0 m c/c in a square pattern	48	65
(3) PVDs (width = 100 mm, t = 4 mm) @ 1.5 m c/c in a square pattern	125	30

Table 2.2 Physical, mechanical, and hydraulic properties of PVD (after [7])

Properties	Quantifier	Specified value
Drains		
Weight per unit length	Minimum	70 gm/m
Width	Minimum	100 mm
Thickness	Minimum	4 mm
Water discharge capacity	Minimum	2840 m^3/year
Core		
Tensile strength	Minimum	750 N
Filter jacket		
Apparent opening size (AOS)	Maximum	90 μm
Grab tensile strength	Minimum	400 N
Elongation at break	Minimum	50%
Puncture resistance	Minimum	130 N
Burst strength	Minimum	800 kPa
Permeability	Minimum	2.0×10^{-4} m/s

for 3 m thick layer and 7 m thick layer, respectively, as mentioned earlier). The times for 90% consolidation with prefabricated vertical drains (PVDs) at 1.0 m and 1.5 m spacings in square pattern were calculated to be 48 days and 125 days, respectively. Approximate costs in the year 2009 for each of the options are also shown in Table 9. Based on the comparison, PVDs (width = 100 mm, $t = 4$ mm) at 1.0 m c/c in a square pattern was considered for implementation by the project owner (Chittagong Port Authority) due to the lowest consolidation time.

Table 2.2 shows the physical, mechanical, and hydraulic properties of the prefabricated vertical drain specified in the design on the basis of PVDs available in the South Asian market. Major design requirements for the PVDs are discharge capacity, tensile strength, and the apparent opening size (AOS). Discharge capacity of the drain should be large enough to ensure efficiency of the drain and to avoid drainage congestion. Reference [12] established that drainage congestion in vertical drain can be ignored if following condition is met.

$$\frac{\pi}{4}\frac{k_{\mathrm{h}}}{q_{\mathrm{w}}}l^2 \leq 0.1 \tag{2.1}$$

For the container yard at Chittagong Port Park area, the required discharge capacity was calculated to be $q_w > 28.0$ m^3/year using Inequality (2.1) and $k_{\mathrm{h}} = 2.3 \times 10^{-9}$ m/s and $l = 7.0$ m. The discharge capacity specified in Table 10, based on the commonly available PVDs, is much greater (100 times) than the required value.

Apparent opening size (AOS) of the filter should be large enough to provide sufficient permeability, yet small enough to prevent the fine particles of the soil from entering the filter and the drain. The permeability of the filter is generally expected to be larger than the permeability of the surrounding soil by at least one order of magnitude from the consideration of clogging effects [5]. Permeability for the PVD was specified to be 2.0×10^{-4} m/s, based on the information of most PVDs available in South Asia. This value is almost 10^5 times greater than the permeability of the soil ($k_h = 2.3 \times 10^{-9}$ m/s).

To prevent the penetration of fine particles, commonly used design criteria are as follows [4]:

$$O_{95} \leq (2 - 3)D_{85} \tag{2.2}$$

$$O_{50} \leq (10 - 12)D_{50} \tag{2.3}$$

where O_{95} is the AOS of filter, O_{50} is the size that is larger than 50% of the fabric pores, and D_{85} and D_{50} are the sizes for 85% and 50% of passing of soil particles by weight, respectively.

Based on field observations of PVDs, a more relaxed criteria for O_{95} was found adequate for Singapore and Bangkok clay [3, 5]. For the silty clay or clayey silt encountered at the site for the container yard in the Chittagong Port Park area, the PVDs design criteria of $O_{95} < 90$ μm were found satisfactory.

PVDs should have adequate strength to sustain the tensile load applied during installation. Therefore the strength of the core, the filter, and the entire drain are generally specified during design of PVDs. [10] suggested that a drain must withstand at least 0.5 kN of tensile force along the longitudinal direction without exceeding 10% in elongation. Tensile strength of the core was specified to be 0.75 kN in the current project. Filter jacket was required to have minimum grab tensile strength, puncture resistance, and burst strength of 400 N, 130 N, and 800 kPa, respectively as shown in Table 2.2.

Surcharge load required for pre-consolidation was estimated to be equivalent to the design load of 56 kPa. Considering unit weight of sand fill as 18 kN/m^3, required height of the surcharge fill was approximately estimated to be 3.0 m.

2.4 Construction Methodology

The construction of the ground improvement work involved preparation of the existing ground, placement of local sand to raise ground level where necessary, placement of a drainage blanket of coarse sand, installation of Prefabricated Vertical Drains (PVDs), and then preloading. Considering the large area (60,700 m^2) of the container yard, the ground improvement works were accomplished in three segments with each segment consisting of approximately 20,200 m^2 of area. Construction methodology of ground improvement works is summarized below:

(i) An approximately 150 mm thick local sand layer was first placed over the leveled ground after stripping topsoil and unsuitable materials in order to attain the required grade for the container yard. The local sand had the fineness modulus (F.M.) of greater than 1.0 and the fine content (material passing sieve No. 200 sieve) of less than 3%. The layer of the local sand was compacted using vibratory rollers to obtain a relative density of at least 85%.

(ii) A drainage blanket consisting of coarse sand (fineness modulus greater than 2.2) was then placed over the local sand to facilitate draining of water to be collected by the PVDs. Thickness of the drainage blanket layer was 450 mm, which was designed to compensate for the settlement expected due to the consolidation. Lower 250 mm of the drainage blanket was placed before installation of the PVDs to provide a working platform for PVD installation. The remainder of the drainage layer was placed after installation of PVDs for allowing the drains to discharge into the sand layer. Surface of the sand blanket was adequately compacted using vibratory rollers and then leveled. The degree of compaction of the sand layer was such that the relative density of the compacted sand is at least 85%.

(iii) PVD was installed using a mandrel that provided minimum sub-soil disturbances. A hollow mandrel or sleeve was advanced through the sub-soil using vibratory, constant load, or constant rate of advance methods. The mandrel combined with the anchor had a maximum projected cross-sectional area of 70 cm^2. The anchor was used to remain in place at the bottom of the PVD when the mandrel was removed after installation. As mentioned earlier, the PVD was installed to the depth of 9 m below ground level to cover the full depth of the soft soil.

(iv) The remaining 200 mm layer of drainage blanket consisting of coarse sand was placed over the finished surface after installation of PVDs which was compacted as per specification and leveled for placement of settlement measuring gauges.

(v) Thirty settlement measuring gauges were placed over the finished surface of the upper 200 mm layer of coarse sand to measure the rate and magnitude of the settlements. A settlement gauge includes a base plate and a stand pipe. The base plate of the gauge was placed on top of the leveled granular layer, while the elevation of the top of the stand pipe was monitored using a Surveyor's level to obtain ground settlements.

A surcharge load consisting of 3.0 m high fill of sand was placed over the 200 mm thick coarse sand layer for preloading. Surcharge material was placed in two layers of approximately equal thickness. Total area was divided into three segments for placement of surcharge and settlement monitoring as well as for installation of PVDs so that material from one segment could be reused in the other segment when consolidation in the first segment was completed as established through monitoring time-settlement responses. The sides of the surcharge load were kept vertical along the boundary of the area using sand bags or brick stacks.

Figure 2.5 shows schematic diagram of a settlement gauge along with the approximate locations of settlement measurements. A schematic diagram showing a profile

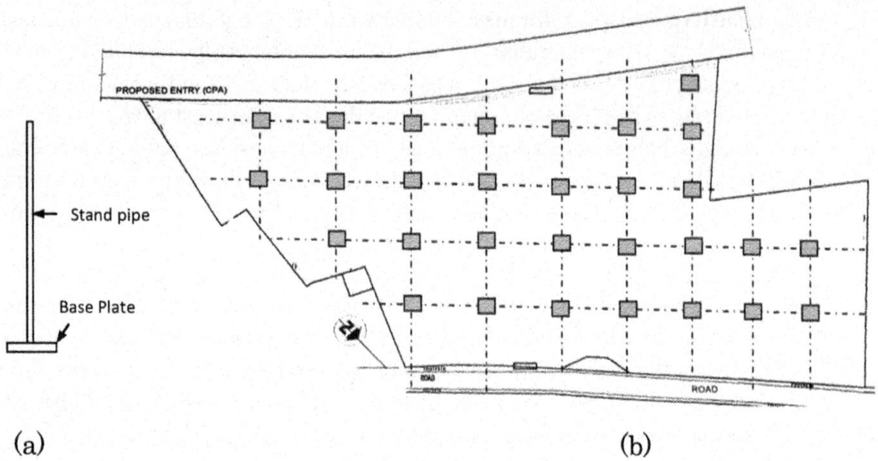

Fig. 2.5 Settlement monitoring scheme **a** Settlement gauge **b** Locations of settlement measurement (after [7])

detailing of the ground improvement works including surcharge, sand blanket, settlement plate, and PVDs is presented in Fig. 2.6.

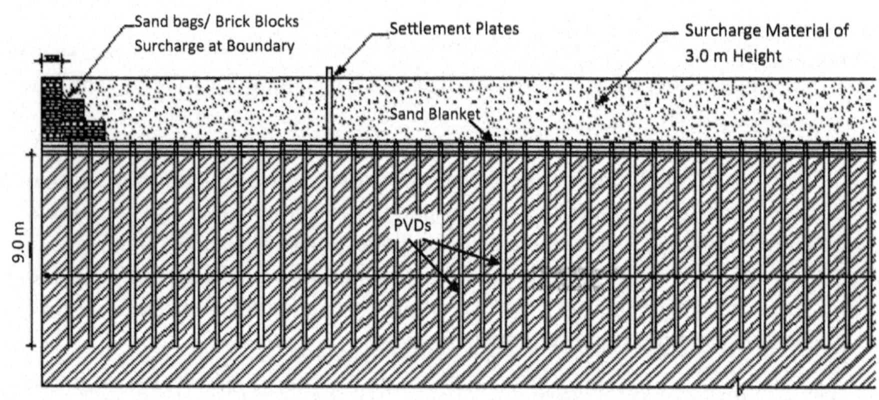

Fig. 2.6 Schematic diagram showing details of ground improvement works (after [7])

2.5 Monitoring of Ground Settlements

Ground settlements were monitored using the settlement gauges during consolidation under surcharge preload to validate the design assumptions and to ensure pre-consolidation before construction of the container yard. Surveyor's leveling apparatus was used to monitor the movement of the tips of the standpipes of the settlement gauge. Settlement monitoring was started immediately after placement of the surcharge to the full height (i.e., 3.0 m) and continued until the consolidation was completed. It generally took 10–12 days to place the surcharge materials to the full height within a segment of the whole area. Immediately after placing the surcharge to the full height within a segment (in 10–12 days), the ground settlements were first measured using the settlement gauges within that segment. The first measured settlement for each gauge is termed herein as the immediate (initial) settlement. The same procedures of surcharge placement and settlement monitoring were then followed for the rest of the container yard area. As mentioned earlier, the whole container yard area was divided into three approximately equal segments for placement of surcharge materials.

Figure 2.7 shows the results of settlement monitoring at different locations within the area of pre-consolidation. Figure 2.7 shows initial settlement of 80–300 mm due to the placement of 3.0 m high surcharge before measurement was started. The maximum settlements measured during the monitoring period varied from 220 to 415 mm, which are very close to the settlements estimated during design stage. As mentioned earlier, the maximum settlements were estimated to range from 203 to 536 mm during design. Thus, the one-dimensional consolidation theory appeared

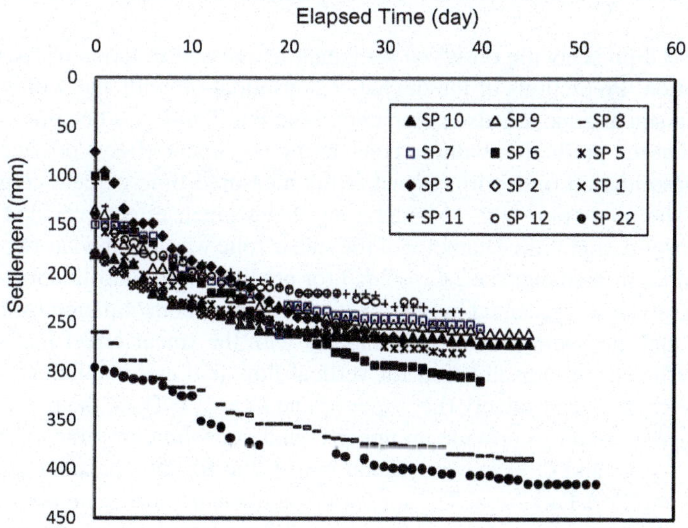

Fig. 2.7 Results of ground settlement monitored with time (reproduced from [7])

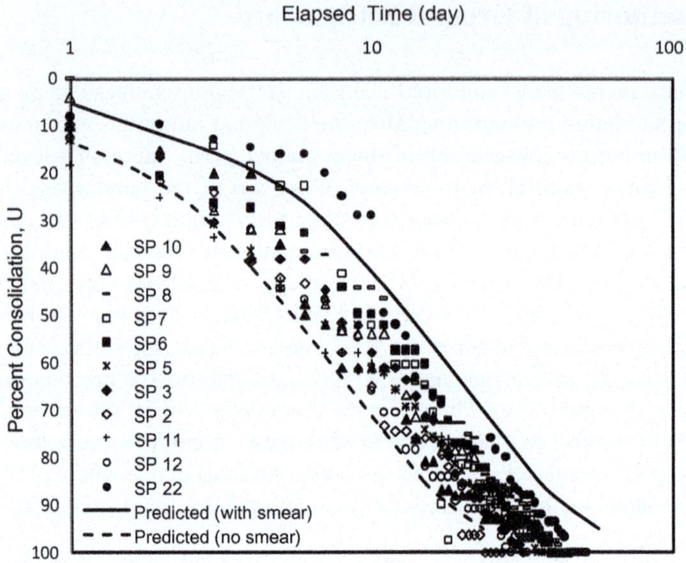

Fig. 2.8 Comparison of measured and predicted degree of consolidation (reproduced from [7])

to reasonably estimate the settlements for the 60,700 m^2 of loaded area overlying 3.0–7.0 m thick layer of compressible soil. Figure 2.7 revealed that the settlements were almost completed after 30–52 days of preloading. Time of consolidation was estimated to be 48 days during design with the PVDs at 1.0 m center-to-center spacing (Table 9). Thus the estimated time reasonably agreed with the observed consolidation time.

Figure 2.8 presents the observed settlement expressed in terms of the degree of consolidation. Predictions of the degree of consolidation with and without considerations of smear zone are also compared in the Fig. 2.8. It has been revealed from Fig. 2.8 that the prediction without considering the smear effects provides a lower bound consolidation time with respect to the measured time-percent consolidation relation. The prediction with consideration of the smear effect provided an upper bound consolidation time. Diameter of the smear zone was assumed as two times the equivalent drain diameter (i.e., $d_s = 2d_w$) for prediction of consolidation time with the smear effects as presented in Fig. 2.8. Coefficient of consolidation for horizontal flow (c_h) and the coefficient of permeability with the smear effect (k_s) was taken as the coefficient of consolidation for vertical flow (c_v) and coefficient of vertical permeability (k_v), respectively (i.e., $c_h = c_v$ and $k_s = k_v$). Thus, the prediction with $d_s = 2d_w$ was found to provide an upper bound consolidation time. The assumptions of $c_h = c_v$ and $k_s = k_v$ were found reasonable for the container yard project at Chittagong Port Park area. The measured settlements during monitoring period confirmed the occurrence of smear, leading to the measured consolidation time to be greater than the predicted value without consideration of smear effects.

2.6 Summary

A comprehensive field and laboratory investigations were carried out for the design of ground improvement for the container yard at Port Park area of Chittagong Port, Bangladesh. The major findings and conclusions based on field investigations, laboratory investigations, design of ground improvement, and field monitoring of the ground improvement work are summarized below:

- Ground settlements and the consolidation time predicted using classical theory of consolidation with parameters obtained from laboratory tests were similar to those observed during field monitoring. It thus appeared that the classical theories can effectively be used in computing the consolidation settlement and the time for consolidation for the 60,700 m^2 container yard area overlying 3.0–7.0 m thick very soft to soft compressible soil.
- Installation of the vertical drains predicted significant reduction in pre-consolidation time (1 year and 5.5 years for 3 m thick and 7 m thick layers of soft soil, respectively without vertical drains to about 50 days with PVDs).
- The Hansbo theory without consideration for smear effects predicted lower bound consolidation time while prediction considering smear effects with smear diameter two times the equivalent drain diameter provided upper bound consolidation time.
- To account for the smear effects, the assumptions for the coefficient of horizontal consolidation and the coefficient of horizontal permeability as those for vertical flow (i.e., $c_h = c_v$ and $k_s = k_v$) was appeared to be satisfactory for design of ground improvement of the container yard project.
- The effect of drainage congestion can generally be neglected in most prefabricated vertical drains with sufficient discharge capacity.

Acknowledgements The successful completion of the project is due to the combined efforts of the project owner (Chittagong Port Authority, Bangladesh), Consultant (Department of Civil Engineering, Bureau of Research, Testing and Consultation, Bangladesh University of Engineering and Technology, Dhaka, Bangladesh), Contractor (Al-Amin Construction Company Limited) and the Sub-contractor. The authors gratefully acknowledge the support and hard work put forth by all field engineers, contractors and sub-contractors, and the workers without whom, the project would not have been successful.

References

1. Ansary, M.A., Daulah, I.U.: Ground improvement—state of the art report and application prospect in Bangladesh. In: Proceedings of the First Bangladesh–Japan Joint Geotechnical Seminar on Ground Improvement 1–25 (1993)
2. Bergado, D.T., Assakami, H., Alfaro, M., Balasubramaniam, A.S.: Smear effects of vertical drains on soft Bangkok Clay. J. Geotechn. Eng. **117**(10), 1509–1529 (1991)
3. Bergado, D.T., Alfaro, M.C., Balasubramaniam, A.S.: Improvement of Soft Bangkok clay using vertical drains. Geotext. Geomembr. **12**(7), 615–663 (1993)

4. Carroll, R.G.: Geotextile filter criteria. Transp. Res. Rec. **916**, 46–53 (1983)
5. Chu, J., Bo, M.W., Choa, V.: Practical considerations for using vertical drains in soil improvement projects. Geotext. Geomembr. **22**, 101–117 (2004)
6. Das, B.M.: Advanced Soil Mechanics. Hemisphere Publishing Corporation, New York (1983)
7. Dhar, A.S., Siddique, A., Ameen, S.F.: Ground improvement using pre-loading with prefabricated vertical drains. Int. J. Geoeng. Case Histories, USA **2**(2), 86–104 (2011)
8. Hansbo, S.: Consolidation of clay with special reference to influence of vertical sand drains. In: Proceedings of Swedish Geotechnical Institute, No. 18 (1960)
9. Hansbo, S.: Consolidation of fine-grained soils by prefabricated drains. In: Proceedings of the 10th International Conference on Soil Mechanics, Stockholm, vol. 3, pp. 677–682 (1981)
10. Kremer, R.R.H.J., Oostven, J.P., Van Weele, A.F., Dejager, W.F.J., Meyvogel, I.J.: The quality of vertical drainage. Proc. Eighth Eur. Conf. Soil Mech. Found. Eng. Helsinki **2**, 721–726 (1983)
11. Siddique, A., Safiullah, A.M.M., Ansary, M.A.: Characteristic features of soft ground engineering in Bangladesh. In: Nakase, A., Tsuchida, A. (eds.). Proceedings of International Symposium—ISYokomama 2000 on Coastal Geotechnical Engineering in Practice, Yokohama, Japan, vol. 2, pp. 231–248 (2002)
12. Xie, K.H.: Consolidation Theories and Optimisation Design for Vertical Drains. Ph.D. Thesis, Zhejiang University, China (1987)
13. Yoshikuni, H., Nakanodo, H.: Consolidation of fine-grained soils by drain well with filter permeability. Soils Found. **14**(2), 35–46 (1974)

Chapter 3
Soil–Cement Method for Jobsite Road Foundation Ground Construction—Using "Moist Cement"

Chie Wakahara and Tomokazu Furuse

3.1 Overview of the Project

The construction project of Tokai-loop Expressway, which connected many cities in and around Nagoya, Japan was planned. The length of the highway was 160 km and its design concept was based on the Standards for Highway Design [1]. Since the elevated type expressway was adopted for this project, many jobsite roads were needed for carrying out the operation of construction machines such as cranes, dump trucks, etc. It was necessary that temporary roads of the jobsites were stable enough for the safe execution of the work, especially when using heavy and mobile construction machines. At many jobsites, the top soils consisted of soft grounds containing clay and silt. Therefore, these soft grounds had to be stabilized by appropriate ground improvement techniques, or by adopting some other countermeasures. After consulting many methods and case histories, the ground improvement techniques by some cement-based stabilizers (soil–cement method) were adopted. By adopting soil–cement, the traffic-ability of the soft ground can be improved and the surplus soils from jobsites can be utilized effectively. However, a major concern was the dusting by cement-based stabilizers during the project execution. This is because most of cement-based stabilizers for soils are in dry powdered state and the project sites were located near the rice fields and agricultural farms.

Although, dust suppression types of cement-based stabilizers are commercially available, however, they are more expensive than general cement-based stabilizers. Considering cement-based stabilizers and their effects, the construct-ability and the economic benefits, etc., soil–cement method by "Moist Cement" was adopted.

C. Wakahara (✉) · T. Furuse
Toyo Sutabi Co., Ltd., Gifu, Japan
e-mail: c-wakahara@toyostb.co.jp

T. Furuse
e-mail: t-furuse@toyostb.co.jp

H. Hazarika et al. (eds.), *Practices and Trends in Ground Improvement Techniques*, Developments in Geotechnical Engineering,
https://doi.org/10.1007/978-981-19-3322-6_3

3.2 Description of "Moist Cement"

3.2.1 Specification of "Moist Cement" Mixer

Soil-cement method is described in Chap. 4 of Part I. Soil–cement is the product of a mixture of pulverized soil and Portland cement and water, and the mixture is compacted to high density. However, this method has one disadvantage: it is not effective for fine-grained soil and organic soil.

In order to make the method suitable for adoption in clayey and organic soils, cement-based stabilizers were developed by adjusting the chemical composition of the cement-based material (for the convenience of description, this cement-based stabilizer will be named simply "cement" in the following discussion).

Preparing flow of "Moist Cement" is shown in Fig. 3.1 [2]. A constant amount of "cement" is supplied from the cement tank to the mixer through the screw feeder. Simultaneously, freshwater, which is controlled quantitatively by flowmeter, is ejected from a nozzle in the mixer. After "cement" and the water are mixed, this mixture turns into "Moist Cement". The amount of water in the mixer can be adjusted in the range of 5–10% of "cement" weight. Figure 3.2 [2] shows a typical mixer for "Moist Cement", which is specified for a trailer-mounted type.

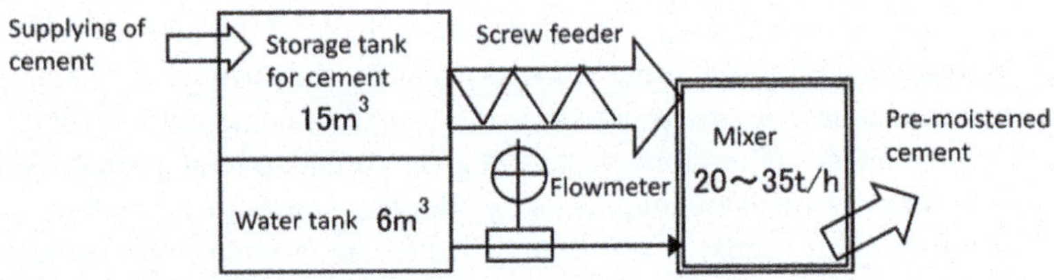

Fig. 3.1 Preparation of the system of "Moist Cement" mixer [2]

Fig. 3.2 Appearance of "Moist Cement" mixer [2]

3.2.2 *Construction Procedure of Soil–Cement with "Moist Cement"*

Execution process of soil–cement with "Moist Cement" can also be achieved by using the conventional powder mixing method for shallow soil layers. Figure 3.3 [3] shows the comparison between the conventional soil–cement method and soil–cement method with "Moist Cement", but the deference of both the methods lies only in producing of "Moist Cement" in the third step.

The execution of soil–cement with "Moist Cement" involves immediate spreading, mixing, and compaction within two hours after preparing "Moist Cement", because "Moist Cement" will lose its solidifying effect after elapse of two hours of its preparation. Generally, the solidifying materials of the conventional soil–cement

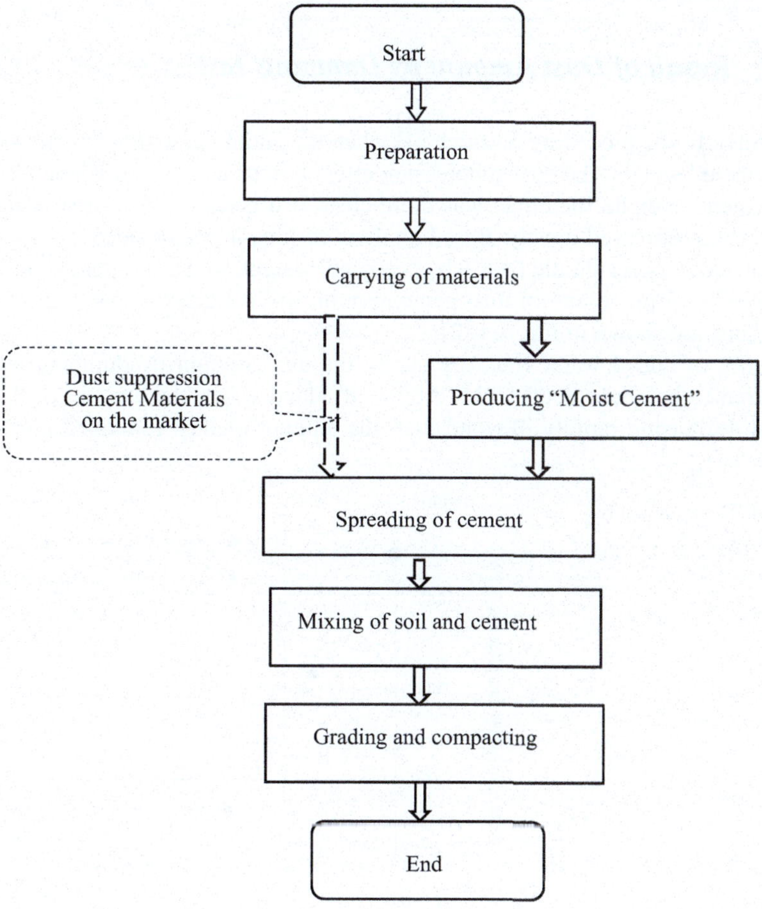

Fig. 3.3 Work flow of soil–cement with "Cement" and "Moist Cement" [3]

method in the jobsite are carried in flexible container bags and they are spread directly from those. However, execution of "Moist Cement" could be carried out with a spreader, a dump truck, or a motor scraper according to the jobsite conditions.

3.3 Properties of "Moist Cement"

"Moist Cement" means the cement-based stabilizer in which some water is added in-situ, and, thus is prepared in wetting state around the surface of solid particles. Consequently, it has the action of dust suppression. However, it is speculated that the strength of mixed soil mass with "Moist Cement" may be lower than with dried "cement". Some of the test results are described below.

3.3.1 Result of Dust Amount by Dumping Test

According to the preceding research [4] when the adding amount of water was 1% by "cement" weight, the dust suppression effect was observed, and when it was 2% by "cement" weight, the effect was more clear. If it became 5%, the generation of dust was not observed mostly. Based on these results, it was decided that the upper limit of added water should be 8% by "cement" weight. The test method conformed to "F033T test procedure of dust generation on soil solidifying material" [5]. The test results are shown in Fig. 3.4 [2].

At 0% of added water (i.e., dry state), the dust amount by dumping test was measured to be about 2900 CPM. At 3% of added water, the amount of dust has started decreasing rapidly. Beyond that, the amount of dust remained at constant

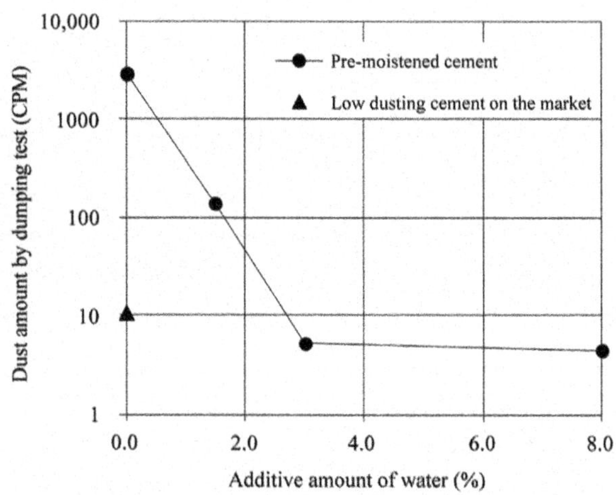

Fig. 3.4 Dust amount by dumping test [2]

value of 5 CPM. This value is less than the dust amount by the dust suppression cement available in the market, which is 10 CPM. Thus, the field values (1–10 CPM) are evaluated in the range of "relative clean room" referring to the generating dust and airborne dust.

3.3.2 Dust Generating State by Trial Work

Based on the above-mentioned test results, the amount of added water to "cement" was decided at 8% by weight, in which the dust suppression effect in the field might be realized. It was evaluated through field observation of the dust suppression state during the spreading and mixing works of "cement". Figure 3.5 [3] shows the results of the trial work. The used materials were the conventional "cement", the dust suppression "cement" available in the market, and "Moist Cement". The trial working stages involve the spreading and the mixing of materials. The dust generations of "Moist Cement" at 8% added water were recognized to be equivalent to those of the dust suppression "cement" available in the market, or less than that.

Fig. 3.5 Dust generation states by trial work [3]

3.3.3 Strength of Solidified Mass

In order to confirm the strength of solidified soil mass by "Moist Cement" with change of water quantities, the strength changes with amount of added water were monitored. Samples of clayey soil and sandy soil were used in which the water quantities were varied up to a maximum value of 8% by weight. Figures 3.6 and 3.7 [3] show the results of the unconfined compression tests of solidified masses. It can be seen that the strengths of solidified mass with 8% "Moist Cement" are about 10% lower than that with non-moistened "cement" and depend on the soil types.

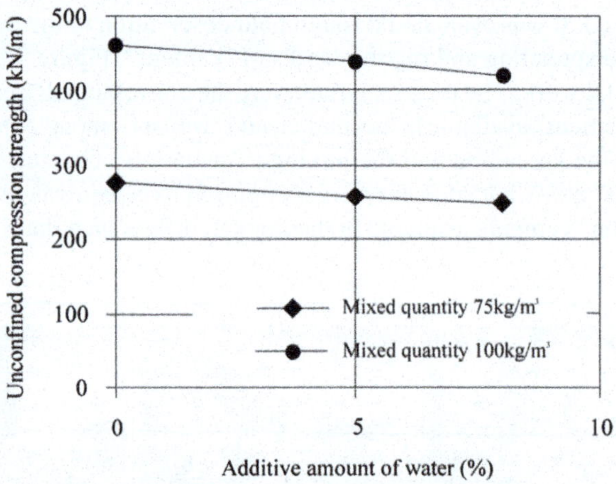

Fig. 3.6 Added water and unconfined compression strength (clayey soil) [3]

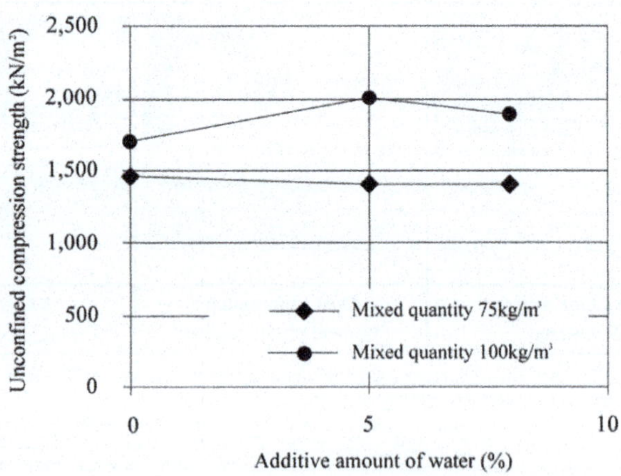

Fig. 3.7 Added water and unconfined compression strength (sandy soil) [3]

3.3.4 Available Time for "Moist Cement"

When water is added to "cement", the hydration reaction of "cement" will gradually take place. To determine the available time of "Moist Cement" after moistening starts, the unconfined compression strengths of solidified masses were measured with the elapsed time after moistening of "cement". The results, with 8% added water, are shown in Figs. 3.8 and 3.9 [3]. From these results, it was found that the strengths of solidified masses decreased 2 h after moistening of "cement".

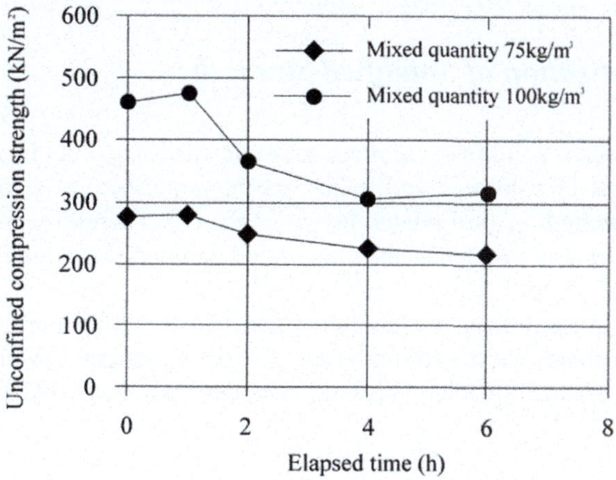

Fig. 3.8 Elapsed time and unconfined compression strength (clayey soil) [3]

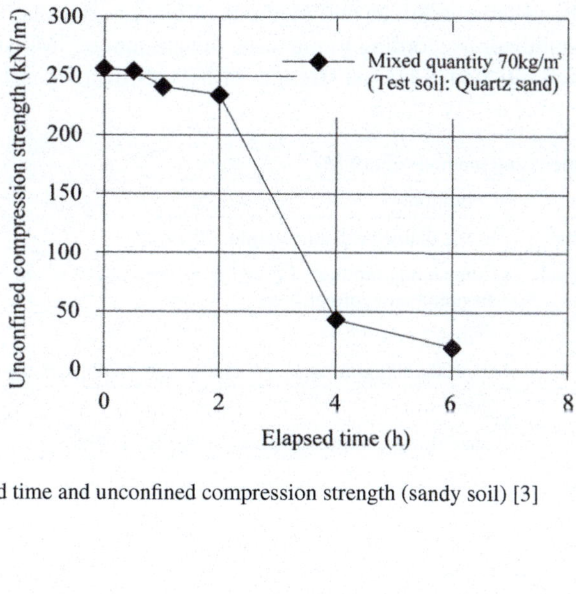

Fig. 3.9 Elapsed time and unconfined compression strength (sandy soil) [3]

3.4 Design Condition

Design strength of the solidified mass is $q_c = 1200$ kN/m^2 derived based on cone index, which is the value of traffic-ability for dump track [6]. Design depth of mixing soil was determined to be 50 cm, which was at a depth with existing soft soils confirmed through the soil survey data.

3.5 Addition Amount of Stabilizers

3.5.1 Investigation of Solidified Strength

The investigation of solidified soil mass strength was carried out by the unconfined compression test of solidified specimens, as the cone indexes q_c and unconfined compression strength q_u, are related by $q_c \fallingdotseq 5q_u$ [7]. Therefore, cone index $q_c =$ 1200 kN/m^2 over was adopted as the unconfined compression strength q_u was more than 240 kN/m^2.

The ratio of strength from the laboratory test and strength at the site was found to be 0.5–0.8 and, therefore, the strength ratio = 0.65 was applied [8]. In this case, the target strength of unconfined compression test was set to be 370 kN/m^2.

3.5.2 Condition and Method of Laboratory Test and Its Result

Table 3.1 [3] shows the conditions and methods of laboratory tests for determining the amount of stabilizer. According to the tests, the amount of "Moist Cement" was 140 kg/m^3, which satisfied the target strength of 370 kN/m^2 (Fig. 3.10) [3].

Table 3.1 Conditions and methods of test [3]

Items	Conditions
Dimension of mold	Inside diameter: 5 cm, Height: 10 cm
Compacting method	Compacting rammer: 1.5 kg, Dropping height: 20 cm 3 layer and 12 dropping number per layer
Stabilizer	"Moist Cement"
Adding amount	50, 100,150 kg/m^3 (per wet unit weight of object soil)
Curing days	28 days
Test	Unconfined compression test (JCAS L-01-2006)

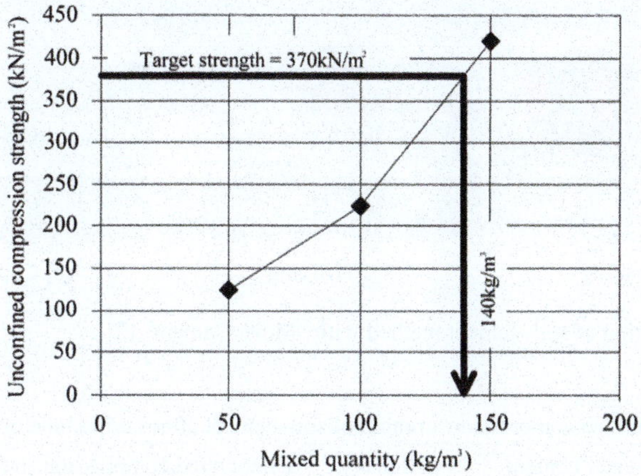

Fig. 3.10 Relationship between the amount of "Moist Cement" mixed and unconfined compression strength [3]

3.6 Execution Specification and Quality Control

3.6.1 Execution Process

The execution procedures at the jobsite are described as below:

(1) Divide the working space in order to perform a uniform spreading.
(2) Supply cement and water to the mixing plant. In this case, the supplying water should be fresh and clean, in order to prevent clogging of a jetting nozzle and flowmeter's troubles due to suspended solids.
(3) Produce "Moist Cement" in the mixer, and load it into the spreader.
(4) Spread it with prescribed amount on the designated space by the spreader.
(5) Mix it with the soil to predetermined depth by the stabilizer, immediately after its spreading.
(6) Level the mixed soil–cement by the bulldozer or the motor grader.
(7) Compact the layer of mixed soil–cement by the tire roller.

Figure 3.11 [2] shows typical working situations at the site.

3.6.2 Execution Management and Quality Control

(1) Certification of improved depth

Certification of mixed and treated depth with soil–cement was carried out by the actual measurement at a drilled point for every 1000 m^2 of the treated area.

(a) Spreading (b) Mixing (c) Compacting

Fig. 3.11 Execution of soil–cement method with "Moist Cement" [2]

Table 3.2 Unconfined compression strength of soil–cement after-curing time of 28 days [3]

Sample No	Water content (%)	Wet density (g/cm^3)	Unconfined compression strength (kN/m^2)
No. 21	28.82	1.858	310
No. 10	32.87	1.78	306
No. 06	27.33	1.699	339
No. 19	27.51	1.747	344
No. 02	33.25	1.79	407
No. 12	26.82	1.739	351

(2) Compaction management

Compaction management was carried out by the proof rolling test with Benkelman's beam in order to find the defective portion. If any defective portion is found, those parts should be re-compacted.

(3) Quality control

Quality control on the strength of soil–cement was carried out by the unconfined compression test with samples that were obtained after its mixing, compacting, and curing. Table 3.2 [3] shows the results. From those results, every measured data was found to satisfy 240 kN/m^2 of the target strength.

3.7 Summary

This report describes the temporally jobsite road construction by the shallow layer improvement with "Moist Cement". When the ground improvement with the powdered "cement" is carried out, the dust by fine particles that may be generated around the jobsite is of great concern. In general, the dust suppression cement-based stabilizer may be used for prevention of dust generations. However, these materials are more expensive than usual cement-based stabilizers.

This case study introduced the system to add proper amount of water to the usual "cement" before mixing it with soils. It is clear that this system is more predominant than the dust suppression "cement" on the market, from the viewpoint of its cost and its dust retardant effect.

References

1. Ministry of Land, Infrastructure, Transport and Tourism of Japan: Standards of Arterial High-Standard Highway, In Japanese (1987)
2. Wakahara, C.: An antiscattering technology for ground stabilization with cement-based material. Geotechn. Eng. Mag. **64** (10) Ser. No. 705, 22–23 (2016b)
3. Wakahara, C.: "MOIST CEMENT METHOD", Committee on Ground Improvement, JSMS, Vol. 302, data (2016a)
4. Japanese Patent No.305788. (UBE KOSAN, Published at 1994), In Japanese
5. Japan Road Association: Manual of Survey and Test for Pavement (IV), "F033T Test Method of Dust Generation by Soil Solidifying Materials", In Japanese (2007)
6. Japan Road Association: Earth Working Manual for Road, In Japanese (2009)
7. The Japanese Geotechnical Society: Japanese Geotechnical Society Standards, Geotechnical and Geo-environmental Investigation Methods (Vol. 1), In Japanese (2013)
8. Japan Road Association: Guide Line of Earth Work and Earth Filling for Road, In Japanese (2010)

Chapter 4
Application Examples of the Sand Compaction Pile Method in the United States

Kenji Harada, Mitsuo Nozu, Kazunori Matsushita, and Yuki Imai

4.1 Introduction

The SCP method was developed as a Japanese original ground improvement method in 1956, more than half a century ago. The conventional vibratory SCP method (trade name: Compozer) improves soft ground in a manner that constructs compaction piles made of sand and gravel in soft ground using the vibration of a vibration hammer. Combining two basic ground improvement principles, namely "compaction" and "drainage", the vibratory SCP method enables one machine to be applicable to all types of ground from sandy soil to cohesive soil ground and thereby having been developed as a typical soft ground improvement method in Japan. In 1995, a non-vibratory SCP method (trade name: SAVE Compozer) that can reduce vibration and noise was developed in order to cope with environmental problems like the adverse effects of noise and vibration on the environments surrounding construction sites. Since then, the method has verified its effectiveness as a liquefaction countermeasure in many large-scale earthquakes in the past [1, 2].

This paper reports the effectiveness of the non-vibratory SCP method to reduce vibration and improving ground in terms of liquefaction prevention and settlement reduction. Such effectiveness was confirmed at the construction site in California to

K. Harada (✉) · Y. Imai
Fudo Tetra Corporation, Tokyo, Japan
e-mail: kenji.harada@fudotetra.co.jp

Y. Imai
e-mail: yuki.imai@fudotetra.co.jp

M. Nozu · K. Matsushita
Fudo Construction, Inc., San Mateo, CA, USA
e-mail: mitsuo.nozu@fudotetra.co.jp

K. Matsushita
e-mail: kazunori.matsushita@fudotetra.co.jp

H. Hazarika et al. (eds.), *Practices and Trends in Ground Improvement Techniques*,
Developments in Geotechnical Engineering,
https://doi.org/10.1007/978-981-19-3322-6_4

which the non-vibratory SCP method was applied as a measure to prevent lateral spreading of the ground due to liquefaction. In addition, this paper discusses about liquefaction judgment and settlement prediction according to U.S. standards of the time.

4.2 Non-vibratory SCP Method

The vibratory SCP method has established a record of total construction length of more than 400,000 km so far but the use of a vibration hammer has caused problems with the adverse effects of vibration and noise on the environments surrounding construction sites. The solution for the problems is the non-vibratory SCP method enabling sand compaction piles to be constructed without vibration and noise through the rotary press-in force of a forced lifting/driving device used in place of a vibration hammer.

Figure 4.1 shows the configuration and names of construction machines used in the non-vibratory SCP method. The construction machines used for the non-vibratory SCP method include an SCP machine as the base, a driving device to rotate the casing, and a hydraulic gear-driven forced lifting/driving device. The weight of the construction machine ensures the reaction force required by the forced lifting/driving device. The rotation of the pinion gear is driven by a hydraulic motor. With this configuration, the vibration and noise can be significantly reduced compared with the vibratory SCP method as shown in Fig. 4.2. Also, a wide range of materials from

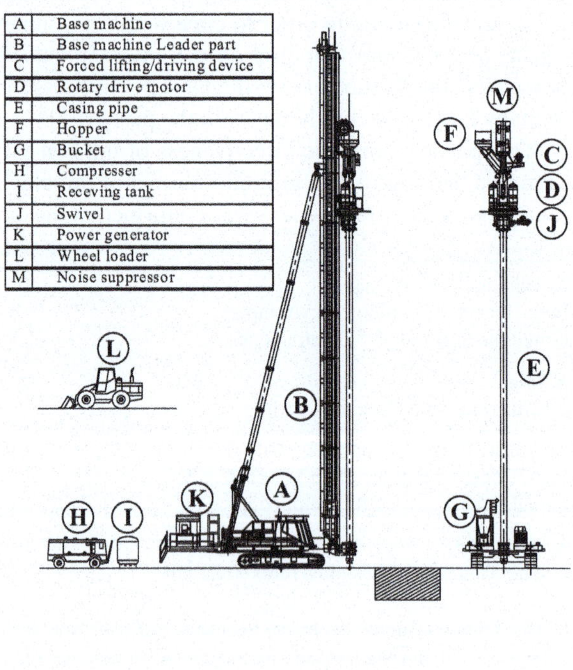

Fig. 4.1 Non-vibratory SCP equipment

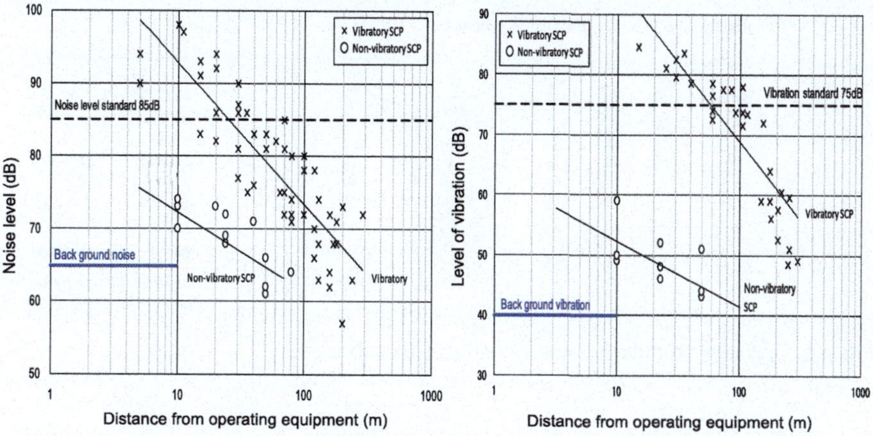

Fig. 4.2 Decrease over distance of noise and vibration

sand to gravel can be used as infilling materials in the SCP method as shown in Fig. 4.3.

The operation procedure of the non-vibratory SCP method is shown in Fig. 4.4. After the casing pipe is rotated and pressed into the ground (① to ③in the figure), it is pulled out while discharging sand into the ground through it (④ in the figure) and rotated as well as pressed back into the ground a short distance (⑤ in the figure) so

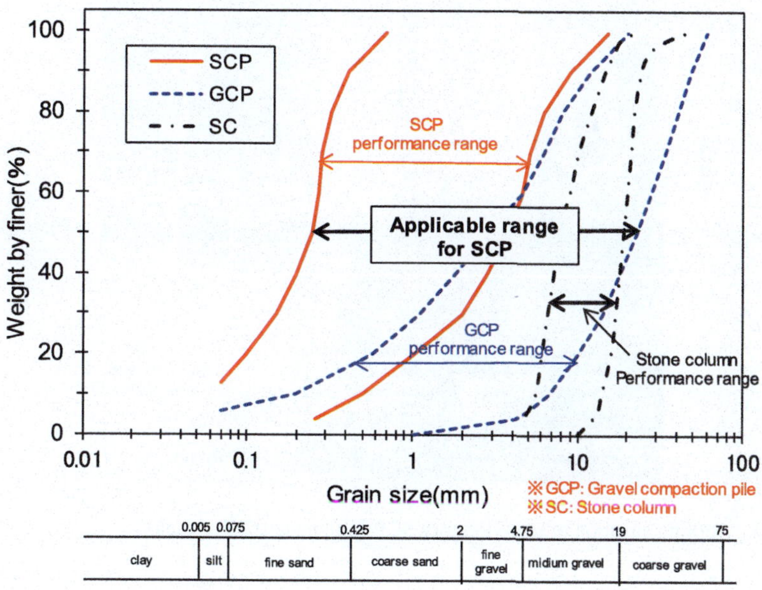

Fig. 4.3 Applicable range of materials

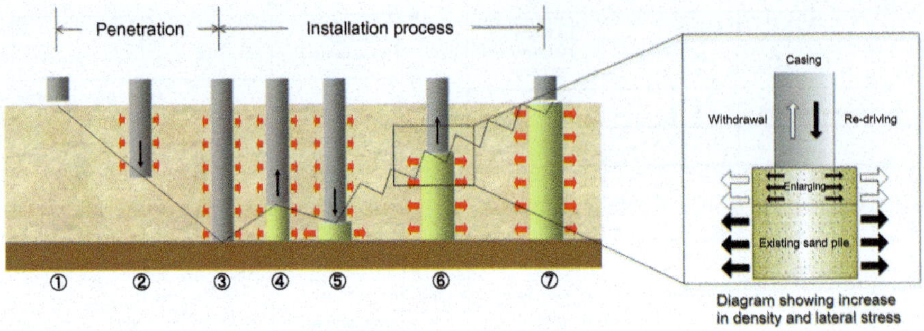

Fig. 4.4 Operating procedure of non-vibratory SCP method

as to enlarge the diameter of the sand pile by repeating the above cycles and thereby compacting the ground around the pile through the press-in effect of the sand pile (⑥ to ⑦ in the figure). Figure 4.5 shows the non-vibratory SCP method operated at the construction site in an urban district in Japan.

Fig. 4.5 Working situation of non-vibratory SCP equipment (Chiba, Japan)

4.3 Example of Applying the SCP Method to a Commercial Facility

This section introduces an example of applying the non-vibratory SCP method to the liquefaction countermeasure on the West Coast of the United States and the results of the vibration monitoring implemented for confirming the vibration reduction effect and the results of pre/post soil surveys [3].

4.3.1 Outline of Project and Construction

The construction was implemented as a countermeasure against seismic lateral spreading of the embankment constructed on reclaimed area to the east of San Francisco Bay shown in Fig. 4.6. The outlines of the project and construction as well as the construction section are shown in Table 4.1 and Fig. 4.7, respectively.

Fig. 4.6 Project location (California State)

Table 4.1 Outline of the projects

Project	Northern California commercial site
Period	February–May 2007
Purpose of improvement	Lateral spread barrier
Application	Non-vibratory SCP Diameter; φ2.5 ft
Quantity	Number of columns = 1642 ea Total liner of improvement = Expansion; 65,306 ft Drain; 686 ft
Criteria	$(q_{c1N})_{cs} \geqq 130$
Specification	Replacement ratio = 12% Diameter of column = φ2.5 ft (expansion) Spacing = 6 ft by 7 ft (square) Expansion area = Top; GL − 3 to − 5 ft Bottom; GL − 19 to − 30 ft

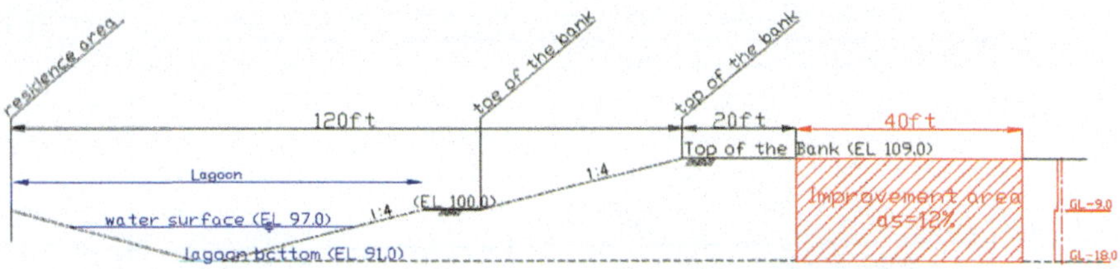

Fig. 4.7 Cross-section of Northern California commercial site

The densification method was planned as the countermeasure for the embankment having a total distance of 28,000 ft (about 8.5 km) along a lake near an upper-class residential area on the opposite shore. The non-vibratory SCP method was decided to adopt in the owner's consideration of the mitigation of the environmental impact on the residential area nearby. The construction site was 40 ft (about 12 m) wide in a transverse direction of the embankment. Also, cone penetration tests (CPT) were conducted for the soil surveys before and after ground improvement. The corrected cone penetration resistance $(q_{c1N})_{cs}$ (Eq. (4.2)) was obtained by correcting the normalized cone penetration resistance q_{c1N} (Eq. 4.1) calculated from cone penetration resistance qc and sleeve resistance fs available through the CPT with a soil behavior type index Ic was required to be more than 130. In order to meet this requirement, the specifications of the non-vibratory SCP method were determined on the basis of the pre-soil survey results as shown in Table 4.1.

$$q_{\mathrm{C1N}} = C_{\mathrm{Q}}(q_{\mathrm{c}}/\mathrm{Pa}) \tag{4.1}$$

where $C_{\mathrm{Q}} = \left(\mathrm{Pa}/\sigma'_{v0}\right)^{n} (n = 0.5)$.

Pa Atmospheric pressure.
σ'_{v0} Effective overburden pressure

$$(q_{\mathrm{C1N}})_{\mathrm{cs}} = \mathrm{Kc} \cdot q_{\mathrm{C1N}} \tag{4.2}$$

where $\mathrm{Kc} = -0.403\mathrm{Ic}^4 + 5.581\mathrm{Ic}^3 - 21.63\mathrm{Ic}^2 + 33.75\mathrm{Ic} - 17.88$.

Ic Soil behavior type index.

4.3.2 Result of Vibration Monitoring During Construction

Vibration was monitored to ensure that the non-vibratory SCP method did not cause vibration affecting the surroundings. As can be seen from the construction situation shown in Fig. 4.8, the site was close to an upper-class residential area. The vibration monitoring to survey the effects of vibration on the residential area was conducted for 7 days, 24 h a day including the time when machines were in operation. Because the levels of vibration of the non-vibratory SCP method are significantly smaller than those of the vibratory SCP method as shown in Fig. 4.9, the construction was reported completed without any complaints.

Fig. 4.8 Working situation of non-vibratory SCP equipment (California, U.S.)

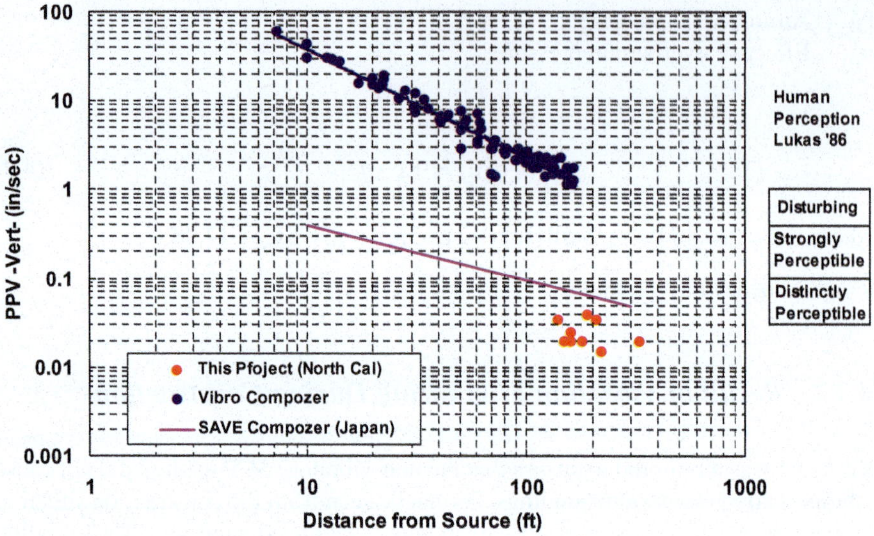

Fig. 4.9 Distance decrement of vibration

4.3.3 *CPT Results*

After the construction, CPTs were conducted at 55 locations in total for confirming the ground improvement effect. Figure 4.10 shows examples of the test results of preconstruction and postconstruction CPTs conducted at almost identical locations. In the figure, the distributions of qc, fs, Ic, and $(q_{c1N})_{cs}$ in a depth direction are obtained through the preconstruction and postconstruction CTPs. As shown in these

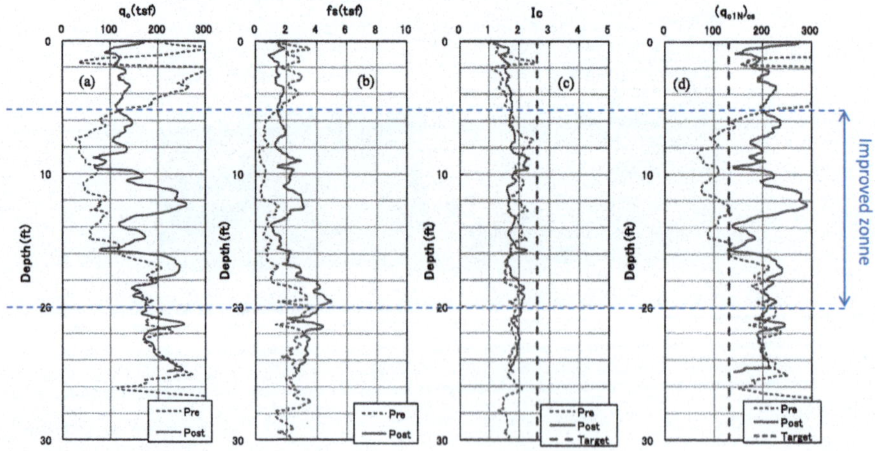

Fig. 4.10 Results of pre/post CPT

test results, it was confirmed that $(q_{c1N})_{cs}$ satisfied the design standard of at least 130 throughout the improvement target range and that the qc values were increased concurrently with the fs values.

4.4 Consideration of the Ground Improvement Effect

This section introduces: the study on the liquefaction judgment and settlement prediction on the basis of the pre/post CPT results described in 3-3; and a discussion on the ground improvement effect of the non-vibratory SCP method.

4.4.1 *Study Conditions*

(a) **The qc and fs values before and after ground improvement**

On the basis of the data shown in Fig. 4.10, with respect to the continuous CPT data at 2 cm intervals in the improvement target range of 5–19 ft (1.5–5.7 m), the averages of the data at 15 points in each 1 foot (about 30 cm) [corresponding to the penetration depth of 30 cm in standard penetration test (SPT)] in the range from 6 to 18 ft (1.8–5.4 m) were obtained so that these averages could be equivalent to the N-values in SPT and used as data for the subsequent study. Figure 4.11 shows the distributions of the averaged values

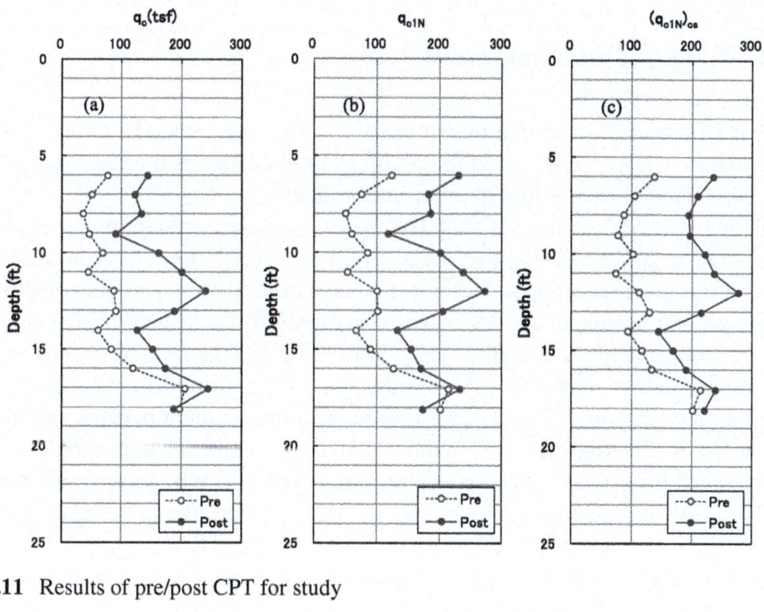

Fig. 4.11 Results of pre/post CPT for study

of qc, fs and q_{c1N}, and $(q_{c1N})_{cs}$ obtained through the pre/post CPTs in the depth distribution.

(b) **Liquefaction judgment and settlement prediction methods used for study**
The liquefaction judgment equation, liquefaction safety factor FS (= CRR/CSS), proposed by [4] was used in this study. The equations to estimate a liquefaction resistance ratio, CRR, and a shear stress ratio, CSR, are also proposed in the paper. In the paper, They describe a method for estimating CRRs using 60% effective *N*-values in the SPT, N_{60} values (used as a standard in Europe and the United States and corresponding to 1/1.3 times the N-values in Japan due to the difference in energy efficiency) and the qc values obtained through CPTs. The prediction method for the settlement following liquefaction proposed by [5] was used in the study. The paper also shows a chart to obtain volumetric strain ε_v from the relation between CSRs and N_{60} values. Thus, in the study, the qc values were converted to the N_{60} values using Eq. (4.3). Also, fines content, Fc values required to obtain CRRs were estimated by using Eq. (4.4). The maximum acceleration ratio α_{max}/g (maximum acceleration at ground surface/gravitational acceleration) used in the design was 0.3 with a magnitude of $M = 7.5$.

$$\frac{(q_c/\mathrm{Pa})}{N_{60}} = 8.5 \cdot \left(1 - \frac{\mathrm{Ic}}{4.6}\right) \tag{4.3}$$

$$\mathrm{Fc} = -1.375\mathrm{Ic}^3 - 3.5 \tag{4.4}$$

4.4.2 Liquefaction Judgment

Figure 4.12 shows the distributions of the N_{60} values calculated from the qc values, the Fc values, CRRs, CSRs, and the FS values on the basis of the results of pre/post CPTs conducted to obtain liquefaction safety factors.

The increase in N_{60} values (Fig. 4.12a) before and after ground improvement was from about 5 to 20. The CRRs (Fig. 4.12c) were about 0.2 before the ground improvement and significantly increased after it. Because the CSRs do not change before and after the ground improvement, the increase in the CRRs caused a significant increase in the FS values (Fig. 4.12d) from less than 1.0 before the ground improvement to much higher than 1.0 after it, which indicates a high ground improvement effect. Regarding the Fc values (Fig. 4.12b), those after improvement became smaller than those before it. The reason for the reduction in the Fc values, which should not have been changed before and after the ground improvement, was pointed out to be the increase in the lateral stress (refer to Fig. 4.4) due to the effect of constructing sand piles unique to the SCP methods. Thus, it can be surmised that the lateral stress ratio Kc (equivalent to a coefficient of earth pressure at rest K_0) was increased.

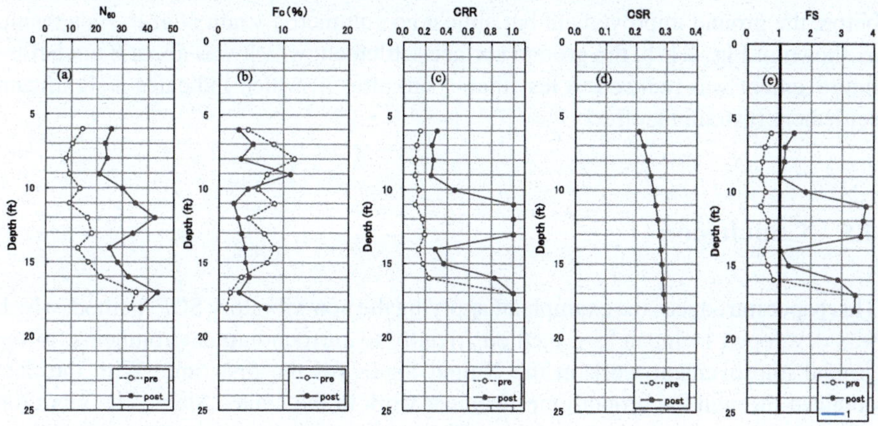

Fig. 4.12 Results of calculation (liquefaction judgment)

4.4.3 Settlement Prediction

Figure 4.13a shows the distributions of volumetric strain ε_v in the depth direction before and after ground improvement. The ground had volumetric strain of 2–3%

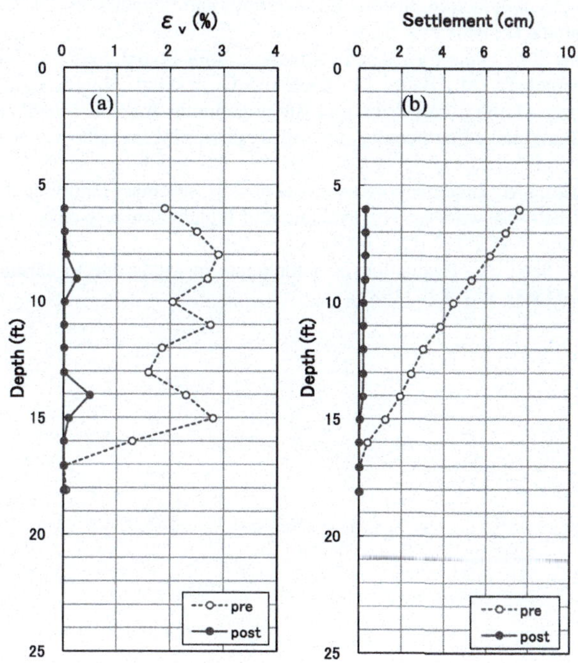

Fig. 4.13 Results of calculation (settlement)

before the ground improvement but almost no volumetric strain after it. As a result, as shown in Fig. 4.13b, the ground surface settlement which was about 8 cm before improvement was reduced to less than 1 cm after it, which indicates a significant settlement restraining effect.

4.5 Conclusion

This paper introduces the example of applying the non-vibratory SCP method, which was developed in Japan as a method friendly to surrounding environments, to the ground improvement work in the United States for the first time. With the data obtained through the ground improvement work in the United States, the vibration restraining effect and the densifying effect of the non-vibratory SCP method were confirmed. The study on the ground improvement effect also showed the increase in the liquefaction safety factor and sufficient settlement restraining effect.

References

1. Yasuda, S., Ishihara, K., Harada, K., Shinkawa, N.: Effect of soil improvement on ground subsidence due to liquefaction. Special Issue of Soils and Foundations, 99–107 (1996)
2. Yasuda, S., Harada, K., Ishikawa, K., Kanemaru, Y.: Characteristics of the liquefaction in Tokyo Bay Area by the 2011 Great East Japan earthquake. Soils Found. **52**(5), 793–810 (2012)
3. Mukai, J., Sakakibara, M., Baez, J.I., Callan, S.G., Tsuboi, H.: First Applications of the SAVE Compozer Method (Non-Vibratory Stone Column) for Soil Densification in U.S.. In: Geotechnical Earthquake Engineering and Soil Dynamics IV, Sacramento CA, Paper 181, 10 pp (2008)
4. Youd, T.L., Idriss, I.M.: Liquefaction resistance of soils; summary report from the 1996 NCEER and 1998 NCEER/NSF workshop on evaluation of liquefaction resistance of soils. J. Geotech. Geo-environ. Eng. ASCE **127** (Oct, 2001)
5. Tokimatsu, K., Seed, B.: Evaluation of settlements in sands due to earthquake shaking. J. Geotech. Eng. **113**(8), 861–878 (1987)

Chapter 5
Cement Deep Mixing for Levee Repair in Louisiana, USA

Sushil Kafle and Ralph Griffin

5.1 Overview of the Project

Hurricane Katrina caused significant damage to the flood control systems in Southeast Louisiana in August 2005. Hurricane Rita further damaged this flood protection system in September 2005. USACE utilized CDM techniques in several phases along the Mississippi River levee to minimize further damage and enhance protection. This project (called P-17A hereafter), which forms a part of the hurricane protection system, is executed under the contract Emergency Levee Repairs, Contract P-17A. This is one of the most significant projects of the USACE after LPV-111. The LPV-111 project (8.9 km of levee treated by CDM) is the largest CDM project ever conducted outside Japan. LPV-111 began in 2010 and was completed in 2012 [1, 2].

The application of CDM at P-17A by installing a series of soil–cement panels oriented in the transverse direction of the existing levee is to provide shear resistance to the levee against the lateral load during high flood water conditions. The project is about 96 km southeast of downtown New Orleans in the Buras district of Louisiana, as shown in Fig. 5.1a. It extends 2224 m along the Mississippi River West Bank Levee, as shown in Fig. 5.1b.

S. Kafle (✉) · R. Griffin
JAFE USA, Inc., San Jose, USA
e-mail: sushilkafle@jafecusa.com

R. Griffin
e-mail: ralphgriffin@jafecusa.com

H. Hazarika et al. (eds.), *Practices and Trends in Ground Improvement Techniques*, Developments in Geotechnical Engineering,
https://doi.org/10.1007/978-981-19-3322-6_5

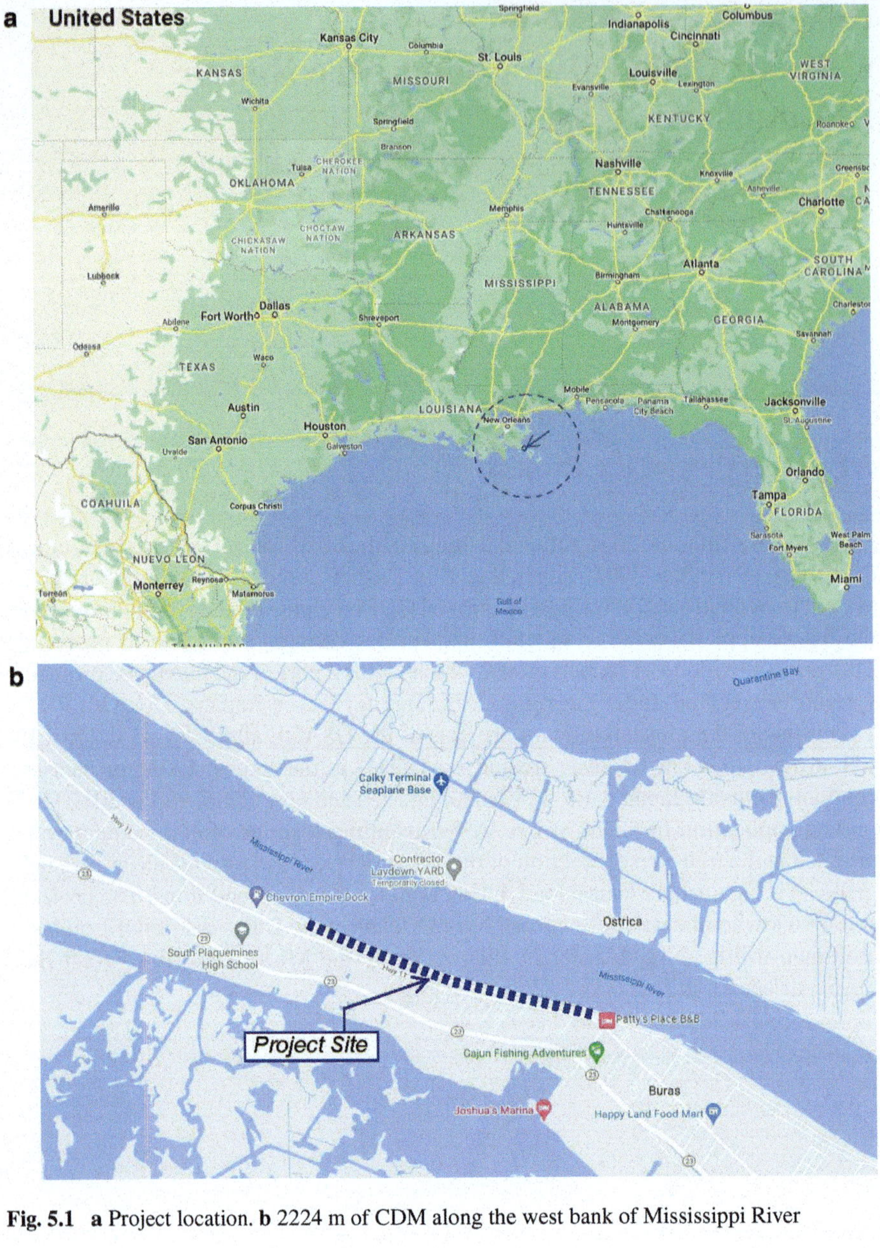

Fig. 5.1 **a** Project location. **b** 2224 m of CDM along the west bank of Mississippi River

Table 5.1 Typical subsurface conditions along P-17A

Soil strata	Properties		
	Moisture content (%)	Unit weight (kN/m^3)	Compressive strength (kPa)
Lean clay (CL)	32–36	18.2–18.4	28–105
Fat clay (CH)	50–70	15.7–18.2	34–120
Silty/clayey sand (SC-SM)	24–25	19.1–19.3	

5.2 Description of Subsurface Conditions

Subsurface investigations work composed of ten borings which were drilled along the levee alignment in January 2014. JAFEC USA Inc. drilled these borings as an additional subsurface investigation to supplement the borings that the USACE completed in the past and to analyze the soil before CDM installation started. Soil samples were obtained from these borings and tested in the laboratory to determine the soil properties. As a result, three primary soil strata identified previously by the USACE were confirmed: lean clay or low plasticity clay (CL), fat clay or high plasticity clay (CH), and clayey/silty sand (SC-SM). Few hints of organic fat clay layers were encountered intermittently. Fortuitously, the extent of organic layers was not significant enough to have a considerable impact. Clay layers were present approximately 8 meters below the ground surface, while sandy layers were widespread below the clay layers. The typical properties of these soil value ranges and ten borings are summarized and listed in Table 5.1.

5.3 Pre-production Testing

Bench-scale testing in the laboratory, as well as full-scale validation tests (pre-production tests), is required before starting the final field production. These validation tests are essential in designing the most cost-effective mixing parameters to ensure uniform mixing and smooth equipment operation in the field.

5.3.1 Laboratory Bench-Scale Testing

Project specifications required the soil mix's field strength (Unconfined Compression Strength, UCS) to be 690 kPa at 28 days. Bench-scale testing (BST) was conducted with preliminary mix designs to investigate the impacts of binder type, binder amount, and water-to-binder ratio on the UCS of the mixes and to obtain a database for the selection of mix designs to be used in the field validation tests. According to soil types (CL, CH, and SC-SM), BSTs were performed using different combinations of binder types, amounts, and proportions. Mainly, two types of binders were included in the test program: Portland cement, PC (100% PC), and Portland blast furnace slag cement, PBFC (25% PC and 75% slag). Binder dosages were 180, 210, 250, and 320 kg/m^3 of soil to be treated, while the water-to-binder ratios (w/b) in the slurry were 80% and 100%. Altogether, 16 batches of mix design were prepared, and a total of 192, 2-inch by 4-inch cylinder samples were cast. These specimens were tested at 3, 7, 14, 28, and 56 days of curing. The 28-day UCS results obtained from these laboratory mixes are shown in Fig. 5.2, which presents UCS results for mixes with PBFC and a w/b ratio of 100%. Mixes with 100% PC and w/b of 80% fulfilled the required strength but were not executed further during the test section; hence, the results are not included in this paper. Figure 5.2 shows that 28-day UCS in the mixes ranged from 1.14 to 4.18 MPa. Also, it can be seen that there is an increase in strength with the increase in the curing time except for a batch where lean clay is treated with 320 kg/m^3 of binder (CL-320). The anomaly in the results for this batch could be attributed to the sampling error during the preparation of samples in the laboratory. There was no significant strength gain for clay soils as the specimens aged from 28 days to 56 days, while the strength almost doubled during the same time interval for sandy soils. The ratio of 56-day UCS to 28-day UCS for clay soils for the binder dosages was constant at around 1.1, while for sand, this ratio was between 1.5 and 1.8.

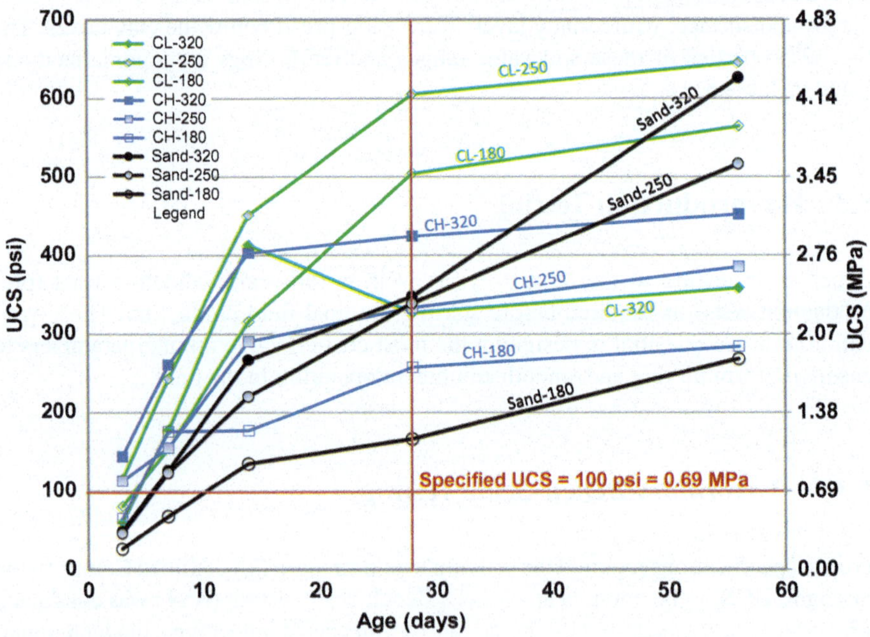

Fig. 5.2 Bench-scale test results. After Shrestha et al. [3]

5.3.2 *Field Validation Test Columns*

The project specifications also required that test columns be installed in three sections about 610 m apart along the levee. Based on the results and analysis of the BSTs, three mix designs were initially proposed for the test sections (180, 210, and 250 kg/m^3 with w/b ratio of 100%). The first set of test columns was successfully installed with these mix designs. However, the workability of the mix proportion of 180 kg/m^3 and 100% was considered too dry from field operations in the remaining two sections. As a result, the w/b ratio in the grout was adjusted to 120 and 150% and maintained within this range for total production depending upon the subsurface conditions. Even though the BSTs gave acceptable strength results in the laboratory, the mix design of 180 kg/m^3 and 100% was not workable in the field. Therefore, field validation tests are crucial in determining the mix designs for total production as only BSTs do not provide a complete apprehension of the in-situ conditions.

5.4 The CDM Equipment

The cement soil mixing process required the CDM rig machine with mixing tools and a grout plant. The plant incorporated two cement silos, a grout mixer, agitator, water tank, and steady-flow pumps. The mixture of cement and water (grout) generated by the plant was pumped to a distance of 1.8 km along the levee to a transfer into an agitator tank from which the grout was, in turn, pumped to the mixing rig. The grout plant could mix more than 160 batches of binder per shift. On average, one batch of grout contains 500 km of the binder. The CDM mixing equipment used in the project was a six-axle auger rig machine, as shown in Fig. 5.3.

5.5 CDM Design and Layout

CDM panels were installed from a working platform (existing top of the levee) and extended from near the top to a competent stratum at depths varying from 7.9 to 14.3 m below the working platform (production pad) as shown in Fig. 5.4. Series of CDM panels, consisting of overlapped soil-cement columns, were installed in a transverse direction to the levee. These transverse CDM panels were designed to enhance the shear resistance of the levee against lateral loads imposed by high water conditions, support additional vertical load created by the increased height of the levee, and mitigate long-term settlement. Geometric requirements such as depth, width, and spacing between the CDM columns and panels were considered when laying out the columns.

CDM panels extended a total length of 2224 m along the levee, and the depths of these panels varied from 7 m in the shallower sections to 14 m in the deeper areas.

Fig. 5.3 CDM mixing equipment

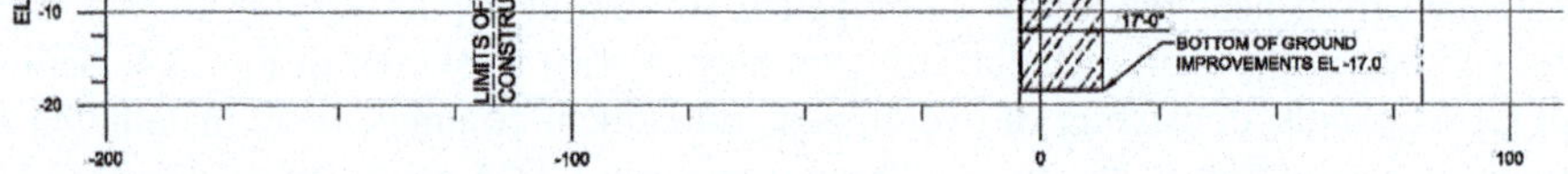

Fig. 5.4 Typical cross-section showing the CDM

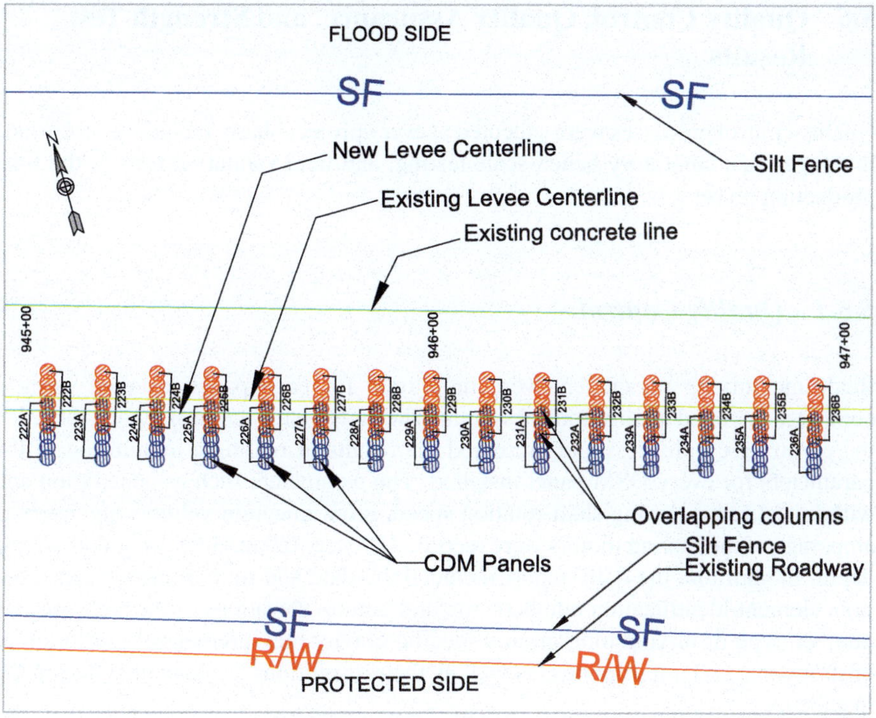

Fig. 5.5 Typical plan view of CDM panels

A typical cross-section showing the existing ground condition and proposed ground improvement is shown in Fig. 5.4, and the plan of the layout design of the CDM panels is shown in Fig. 5.5.

The width of these panels varied from 5.2 m in the shallower section to 10.7 m in the deeper section. To satisfy the width requirement, CDM panels in the narrower section included two elements (one element consisting of six columns) where the three edge columns of each element entirely overlapped at the center (Fig. 5.5). The wider section included up to five elements partially overlapped with the consecutive element installed. The center-to-center spacing between the panels was designed at 3.2 m. A total of CDM installed was 1822 elements of approximately 41,000 m^3 of ground improvement.

5.6 Quality Control, Quality Assurance, and Strength Test Results

Quality control processes were executed at every project phase, including subsurface investigations, laboratory bench-scale testing, and field validation tests to the final production phase.

5.6.1 Quality Control

Real-time mixing parameters were monitored for each panel installed using a computer-operated quality control system. The daily quality control reports recorded the construction progress and presented the results of real-time monitoring of the parameters for every DSM panel installed. The parameters such as penetration and withdrawal rates, mixing shaft rotation speed, grout injection volume, and average amperage of the mixing motors were recorded at every 0.9 m (3 ft) interval along the depth. In addition, the daily report included the date and time of element installation, element identification numbers, top and bottom elevations of the elements, rig number, type of mixing tool, grout's specific gravity measurements, description of obstructions, and interruptions, as well as their resolutions implemented during the installation.

5.6.2 Quality Assurance and Strength Test Results

The production quality was verified for the CDM strength by coring over three percent of the total production elements. Specifications required that these cores be obtained for the full depth of the element and tested at 28 days after installation. A core barrel with an outer diameter of 85 mm and an inner diameter of 63.5 mm was used to retrieve the core samples. Each core run was 63.5 mm in diameter and 1 m long. Cores retrieved along the entire depth were then cut into cylindrical samples 63.5 mm in diameter and 127 mm in height for compressive strength testing, as shown in Fig. 5.6. The minimum core recovery required from the specification was 80%, and an average of 97.8% of the cores was recovered throughout the project.

Fifty-five full-depth cores were taken at approximately every 33 m intervals along the levee. The number of core runs at each location varied depending on the depth of the element. Each 1 m core run was retrieved, logged, and photographed. Two representative samples were prepared from each core run, and one sample was selected for the core run's strength test. The second sample was stored as a backup for the future strength test if required. Over 1100 samples were prepared throughout the project, and 557 samples were tested at 28 days. All results were tracked and analyzed as a

Fig. 5.6 Cored samples

quality control measure to determine that every core met the specification requirements. Figure 5.7 shows an overall distribution of UCS of the core samples obtained along the levee profile from the start station to the end station of CDM. UCS results of all the core samples are included in the plot. Six out of the 557 samples resulted in UCS less than 690 kPa. These samples were retested using backup samples from the same core run, which exceeded 690 kPa. This implied that the quality of test specimens, such as small lumps or invisible micro-cracks, could have influenced the test data.

Project specifications allow that not more than 10% of the tested core samples could fall below the 690 kPa requirement. It can be seen from the frequency distribution curve of the 28-day UCS for the 557 samples tested presented in Fig. 5.8 that

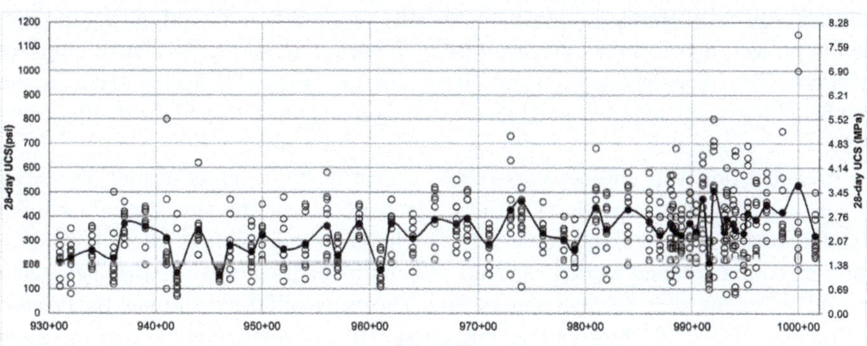

Fig. 5.7 Average 28-day UCS of the cored samples. After Shrestha et al. [3]

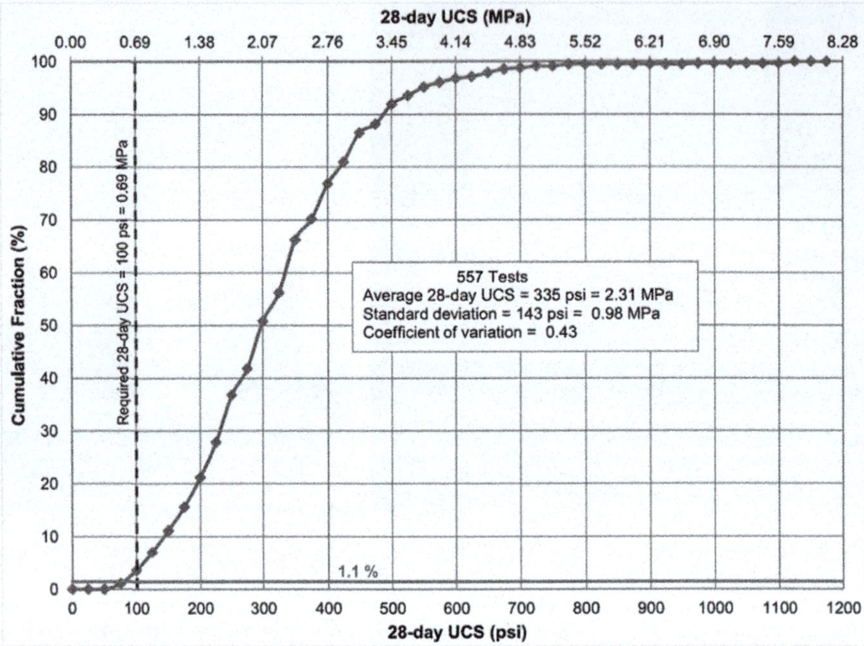

Fig. 5.8 Cumulative frequency distribution of the 28-day UCS result. After Shrestha et al. [3]

1.1% of the tested samples fell below 690 kPa. None of them fell below 551 kPa.

5.7 Construction Challenges

One of the biggest challenges was the weather, among many logistical, technical, and scheduling challenges to face and overcome during the CDM installation. The overall climate in Buras, Louisiana, is usually hot and humid, and during the time of the project, temperatures averaged 92.5 °F during the day, and relative humidity averaged 78%. Special attention and safety measures were established to ensure that the heat illness did not pose a real threat to the on-site workers and the quality of the mix.

Frequent rain was another remarkable aspect of the weather. Rainy days were more than 50% of the production days. On the construction site, the rain quickly created sticky mud, making work very challenging and creating safety hazards in moving equipment and workers around. When dry, pathways were easily navigated, but when it rains, they became untraceable. The rain itself did not hinder the production and quality of CDM columns, but safety and environmental hazards increased dramatically. The rain was often accompanied by lightning, so the lightning strikes in the area were monitored regularly. If lightning was detected within a certain radius of the site, work was shut down as orderly as possible, and workers sheltered until the danger of lightning passed to a safe distance. In general, storms in the area create delays of about 30 min in each occurrence.

Other distinctive safety issues encountered were caused by local fauna. Several stinging insects were found on site. Also, workers were cautioned to be constantly aware of the possibility of alligators in the ditches and edges of the river. Though the site is relatively remote from larger metropolitan areas such as New Orleans, good emergency services were available immediately adjacent to the site. The weather was extremely challenging to the equipment, and the site's remoteness meant that sudden breakdowns could lead to several hours of waiting, while parts and suppliers were procured and brought to the site.

Due to the site conditions and restrictions, the batch plant must be placed at only one location for the entire project's duration. Grout was pumped several thousand feet to an agitator tank of a sub-plant station. Then the collected grout was pumped from the sub-plant station to the CDM rig machine through special pumps. Such long distances created additional time and materials loss during start-up and shut down each day, as well as during unpredicted downtime.

5.8 Summary

With real-time monitoring of the installation parameters, geometric data from field surveys, and full-depth coring and testing data, it was concluded that the CDM elements constructed met the geometric and strength requirements specified for the project. The validation tests led like a guide to adjust the water-to-binder ratio and determine the final mix design. The BSTs alone could not represent the in-situ water need. The overall core recovery of CDM obtained was between 95 and 100%. This good core recovery can be attributed to overall uniformity in mixing and conservation of the strength design.

The CDM operation, quality control, and quality assurance procedures utilized by all parties involved at various phases ensured the project's timely completion, regardless of several production challenges. Valuable lessons learned from the site-specific construction challenges leave us a good message for further improvement in our technology and methods.

Acknowledgements The project could not be successfully completed without the combined efforts of all the project teams, including the owner, designers, contractors, and the sub-contractors. The authors would like to recognize the hard work put forth by all field engineers, superintendents, the general contractors (Clark Construction Group and Hamp's Construction LLC), and the field workers without whom the project would not have been efficiently carried out.

References

1. Bruce, D.A., Cali, P. Woodward, M.: The history of deep mixing in New Orleans. In: 4th International Conference on Grouting and Deep Mixing, Deep Foundations Institute (DFI), New Orleans, LA, February, pp. 15–18 (2012)
2. Cali, P., Lelong, B., Bruce, D., Valagussa, S., Beckerle, J., Gardner, J., Filz, G.: Overview of deep mixing at levee LPV 111. In: 4th International Conference on Grouting and Deep Mixing, DFI (2012)
3. Shrestha, R., Griffin, R., Cali, P., Kafle, S., Watanabe, M., Yang, D.: Deep mixing for levee repair at hurricane protection project, P-17A, Louisiana, USA. In: Proceedings of the 6th International Conference of Deep Mixing, Deep Foundations Institute (DFI), San Francisco, CA, pp. 257–266, 2–5 June 2015 (2015)

Chapter 6
Full-Scale Embankment in Soft Bangkok Clay Using Jet Grouted Cement Mixing Piles

Dennes T. Bergado, Salisa Chaiyaput, and Pitthaya Jamsawang

6.1 Overview

One of the ground improvement schemes that have been adapted is chemical stabilization using cement or lime. In Asia, however, the use of cement has permanently been outpacing the use of lime, not only because cement is abundant and cheaper but also because cement is more effective in terms of strength improvement than lime [3–5]. Ground improvement by cement stabilization can be broadly divided into shallow stabilization and deep stabilization. Shallow stabilization includes stabilization of subgrade for roadways and airfields and other similar structures; it normally employs "low water content" mixing. The deep stabilization, on the other hand, includes deep mixing method (DMM) using either slurries or powder of cement to form columns of improved soil in the ground. Slurry cement mixing is more environmental-friendly as it avoids contact with carbon dioxide which can cause carbonation and health hazards. The improved column of soil is considered to act as reinforcement or as a pile, transferring the load to the skin and to the bottom end of the improved column of soil.

The methods of mixing generally applied in the installation of DMM piles in soft clay deposits are either mechanical mixing or high-pressure jet mixing/grouting

D. T. Bergado (✉)
Asian Institute of Technology, Bangkok, Thailand
e-mail: dbergado@gmail.com

S. Chaiyaput
School of Engineering, King Mongkut's Institute of Technology Ladkrabang, Bangkok 10520, Thailand
e-mail: salisa.ch@kmitl.ac.th

P. Jamsawang
Department of Civil Engineering, King Mongkut's University of Technology North, Bangkok, Thailand
e-mail: pitthaya.j@eng.kmutnb.ac.th

H. Hazarika et al. (eds.), *Practices and Trends in Ground Improvement Techniques*, Developments in Geotechnical Engineering,
https://doi.org/10.1007/978-981-19-3322-6_6

[6]. In the mechanical mixing, the chemical admixtures are mixed into the soil by mixing blades, while in the jet mixing, the chemical admixtures are mixed into the soil through jet of water or slurries of admixtures. Thus, the DMM and jet mixing methods would normally produce high water content cement-admixed clay; besides, the soft clay deposit normally has high water content. High water content cement admixed is hereinafter defined as those clay-cement mixtures having total clay water content (inclusive of water from cement slurry) of at least the liquid limit of the base clay. Moreover, the DMM was applied to mitigate effect of liquefaction [11, 13].

Different names have been used for DMM technology, depending on the stabilizing agent and installation process used. Porbaha [8, 9] suggested adopting a consistent terminology to avoid confusion: "deep mixing." Also, since the terms "soil–cement pile" or "cement pile" or "deep mixing piles" and the terms "soil–cement column" or "deep mixing column" are used interchangeably in the literature to designate the improved column of soil by deep mixing, the term "deep mixing pile" or "deep mixing column" will be used interchangeably in this manuscript to designate the column of improved soil formed due to deep mixing.

An engineered composite material consisting of backfill soil and tensile reinforcement materials commonly known as reinforced soil/earth or Mechanically Stabilized Earth (MSE) has been used extensively for construction of embankments and retaining walls worldwide. In the present study, backfill of MSE wall was developed as a lightweight material with the increasing MSE stability.

Moreover, reinforced embankments were formed by different combinations of tensile reinforcements and facing elements over the years. The most commonly and aesthetically pleasing facing element used in the reinforced embankment construction is the precast concrete facing panel besides other alternatives such as geotextile wrap-around facing, steel plate facing, gabion facing, cast-in-situ concrete facing, welded wire facing, segmental block facing and, etc. Various types of reinforcements commonly used are geotextile, geogrid, steel strip, steel rod, welded wire grid, hexagonal wire mesh, etc. All these components influence the load distribution along the reinforcement in reinforced wall due to different stiffness modulus. Determination of quantity of reinforcement used in reinforced embankment is closely related to the actual tensile load that needs to be stabilized. Therefore, prediction using an accurate method to obtain an optimum quantity of reinforcement is essential.

The performance and behavior of reinforced embankments were monitored and studied accurately by using full-scale test. It consumed a lot of time to construct and involved tedious work during the construction and field monitoring works. Furthermore, it is not cost-effective to construct extra full-scale model and study the performance and behavior of reinforced embankments in order to gain more information when parameters are changed. Numerical simulations by using the finite element model can be one alternative for providing a cost-effective solution to understand behavior of reinforced embankments.

In this case study, soil–cement pile was adopted to improve the soft clay foundation of full-scale embankment test with hexagonal wire mesh reinforcement. In the stability analysis of improved ground, it is normally assumed that reinforced embankment loading will be transferred through the DMM pile down to a firm stratum.

Consequently, the performance of the reinforced embankment and the characteristics of the improved soft soil foundation have to be considered. Lateral movement of reinforced embankment is an essential aspect to be considered as a result of foundation settlement or lateral deformation of foundation sub-soil. Due to the difference in compressibility between soil–cement pile and the adjacent soft soil, complex vertical stress distribution occurs across the foundation level of the reinforced embankment. DMM piles with higher stiffness than surrounding soft soil may cause arching effect within the reinforced embankment. This will attract more vertical stress directly onto the DMM pile. The reinforcement within the reinforced embankment will distribute the remaining vertical stresses and thus minimize settlement.

6.2 Full-Scale Cement Deep Mixing Improved Soft Clay Foundation

6.2.1 Reinforced Embankment

The test location of the field test embankment is located at Wangnoi Power Plant site, Ayuthaya, Thailand. The plan layout for the test embankment is 19 m long and 6 m in width. The dimension of the vertical embankment face made of precast concrete discrete facing panel was 6 m in height with total vertical surface area of 120 m^2. The details of wall geometry are illustrated in Fig. 6.1.

The embankment was made of well-compacted silty sand backfill reinforced with PVC-coated hexagonal wire mesh. The backfill soil has compacted unit weight of

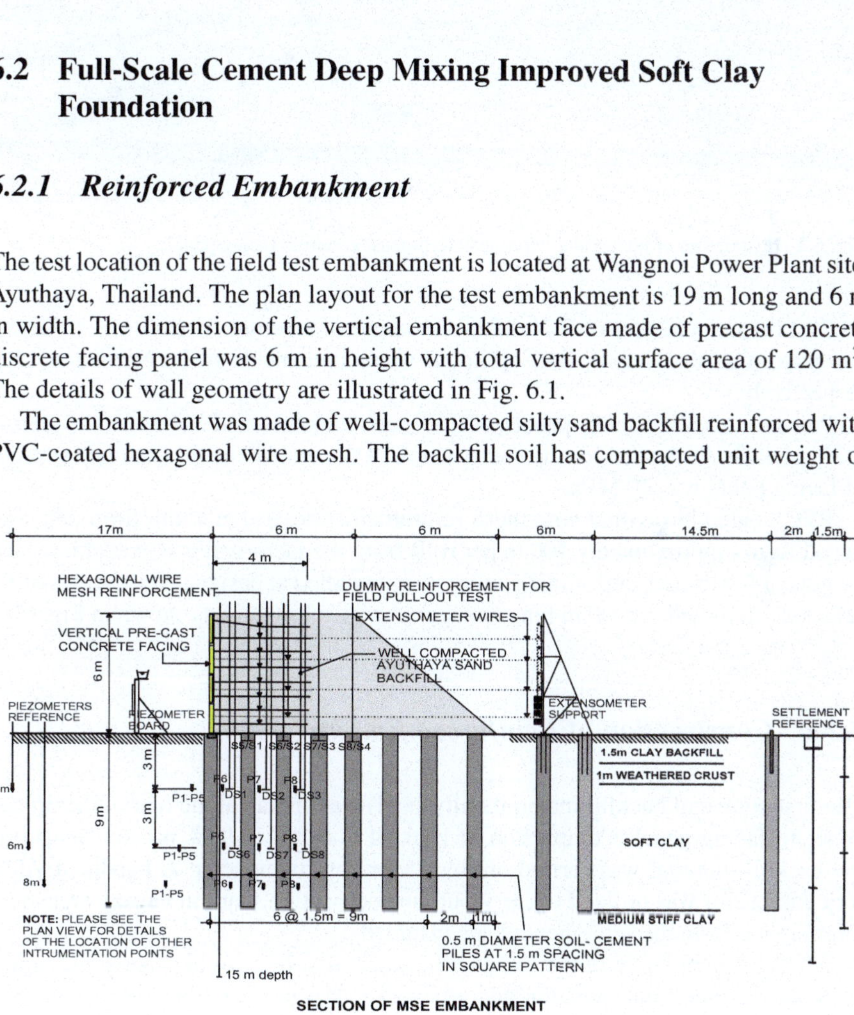

Fig. 6.1 Full schematic diagram of instrumentation at test embankment/wall

Fig. 6.2 Installation of hexagonal wire mesh to precast concrete facing panel

18.20 kN/m^3, cohesion of 7.70 kPa, and angle of internal friction of 22°, and it has maximum dry density and optimum water content of 16.1 kN/m^3 and 15%, respectively.

Precast concrete facing panels of 150 mm thickness and with surface area of 2.25 m^2 each, were used for the construction of the embankment. Concrete strength of facing panel was 30 MPa.

PVC-coated hexagonal wire mesh reinforcement of 1.0 m width form delivered in roll form (approximately 100 m per roll) from the factory. It was precut into strip lengths of 5.0 m each and. strain gauges were installed at designated locations of the selected 1.0 m wide × 5.0 m long hexagonal wire mesh strip as shown in Fig. 6.2.

6.2.2 Construction of Reinforced Embankment

The first course of backfill material (silty sand) was spread at the rear of the precast concrete facing panel. Compaction of backfill material to 95% proctor density of the backfill material was carried out by a vibratory compactor. A handheld vibratory compactor was utilized to carry out the compaction work in narrow areas with interference from the settlement plates installed.

After the completion of the compaction work, first layer of hexagonal wire mesh strip was installed into position and attached to the precast concrete facing panel by a 1.0 m lap joint mechanism. The effective length of the hexagonal wire mesh strip was 4.0 m after deducting 1.0 m for the overlap joint mechanism. The lap joint mechanism was secured by inserting a 12 mm diameter deformed steel bar across the lap fold mesh.

Hexagonal wire mesh strip was installed horizontally in a discontinuous manner at 1.5 m spacing from the center line of the hexagonal wire mesh strip to the adjacent center line of the hexagonal wire mesh strip along the embankment length. Prior to the backfilling operation, hexagonal wire mesh strip installed was stretched and secured into position with some pin device to minimize slack within the hexagonal wire mesh strip.

Subsequent course of backfill material measuring of 0.375 m thick was spread over the plan surface area covering the embankment. Similar compaction work was carried out before the next course of precast concrete facing panels was installed. The procedure mentioned above was repeated until the full height of 6.0 m is achieved.

During the construction of the embankment, field density test at various selected places was carried out using sand cone replacement method to ensure compaction was carried out to minimum of 95% standard proctor density.

6.2.3 Installation of Deep Mixing Piles by Jet Grouting

The foundation sub-soil was first improved with soil–cement piles, called DMM piles, which were installed in-situ by jet mixing method employing a jet pressure of 20 MPa. The water-cement ratio (W/C) of the cement slurry and the cement content employed for the construction of DMM piles were 1.5 and 150 kg/m^3 of soil, respectively.

The DMM piles were installed at 1.5 m spacing in square pattern, except for the perimeter soil–cement piles which were installed at 2.0 m spacing. Each DMM pile has diameter of improvement of 0.5 m and length of 9.0 m, which is up to the bottom of the soft clay layer.

The DMM piles were allowed to cure and the dissipation of excess pore water pressure was monitored until about 80 days before the embankment was constructed.

6.2.4 Sub-soil Profile

The site was underlain by the well-known soft Bangkok clay. The soil profile had a newly-placed 1.5 m thick clay backfill since it was situated within a reclaimed parcel of land. The soft clay, which is overlain by 1.0 m thick weathered crust and 1.5 m thick clay backfill, was encountered from 2.5 to 9.0 m depth. The undrained shear strength obtained from field vane test of the soft clay was less than 15 kPa. The liquid limit and the water content were 100 and 80, respectively. Underlying the soft clay layer is medium to stiff clay layer, having strength of more than 50 kPa.

6.2.5 Instrumentations and Monitoring

The embankment and the improved foundation were instrumented. Piezometers (P), which monitored the dissipation of excess pore water pressures in the foundation soils during and after deep mixing (jet grouting), were installed at various points underground within and outside the embankment zone. Surface settlement plates were installed both "on pile" (S2) and "on clay" (S6) at the bottom of embankments. Deep settlement plates (DS) were also installed at 3.0 and 6.0 m depths at few locations as shown in Fig. 6.1. In addition, vertical (V) and horizontal (H) inclinometers were placed near the vertical side of the embankment to measure the lateral displacement and settlement profile, respectively. The embankment was demolished and all monitoring was eventually terminated after 12 months.

6.3 Numerical Simulation of Full-Scale Test Embankment

6.3.1 Finite Element Model

The finite element model (FEM) was applied to simulate the behavior of field test embankment. Moreover, the simulation results from FEM were compared with the mearing data from instrumentation for confirming the behaviors of embankment.

Two-dimensional (2D) problem for FEM analysis, the PLAXIS software allows to model the test embankment as a plane strain idealization including simulation of construction sequence.

Moreover, the 3D FEM in the PLAXIS 3D Tunnel Program is created through 2D in vertical cross-section model in $x - y$ plane and later extended into the third dimension (z direction).

Under the staged construction feature of the software, the program allowed for the incremental fill placement to be simulated. The zone near the wall face and the top layer of the sub-soil, high-stress gradients were considered and the mesh was refined in these zones. Furthermore, mesh refinement was also applied in-between DMM piles converted walls to assure better demonstrations of differential settlements between untreated ground and DMM piles.

The in-situ stresses in the foundation soil were generated by the K_0 procedure. The backfill which was divided into 10 layers was placed and compacted on the DMM piles improved ground layer by layer. After the placement of a compacted fill layer, the reinforcement in strip form arrangement (not in continuous form of arrangement) was placed on top of the layer of compacted backfill before the next layer was added.

The compacted fill in a given layer was assigned the material parameters according to the stress state induced after the addition of the layer. During this construction stage, undrained analysis was used to simulate the layer-by-layer construction. After the completion of the full height of the embankment, drained analysis was used to simulate the consolidation process.

6.3.2 Boundary and Initial Condition

Standard fixities were used as boundary conditions by fixed boundaries in x, y, and z direction at the bottom of foundation soil. Lateral movement was fixed while vertical movement was allowed at the vertical plane of the foundation soil. The initial stresses in model were influenced by the weight of the material and the history of its formation. Initial stress condition was generated using K_0 procedure built in the program by the coefficient of lateral earth pressure. The stages of construction applied in 3D numerical analyses were parallel to the construction applied in the 2D numerical analyses.

6.3.3 Material Models

Foundation soil, the foundation soil was divided into eight layers and five layers in the 2D and 3D models, respectively. Backfill clayey soil from the ground surface to 1.5 m depth was modeled with soft soil model. Weathered crust soil was at 1.5–2.5 m depth and this material was modeled with Mohr–Coulomb model. From 2.5 to 9.0 m, the presence of this soft clay layer was modeled with soft soil model. Medium stiff clay was laid from 9.0 to 11 m while stiff clay was laid from 11.0 to 20.0 m. Mohr–Coulomb model was used to model these medium stiff to stiff clay layers. Detailed parameters used for these models are shown in Tables 6.1 and 6.2 for 2D and 3D models, respectively.

Soil–cement column, Mohr–Coulomb model was selected to model the DMM clay-cement pile.

Backfill material, the backfill material used in the embankment construction was silty sand sourced from Ayutthaya. The elastic, perfectly plastic Mohr–Coulomb model was used to model this material. The apparent cohesion, c', was 10 kPa and the frictional angle, ϕ', was 30°.

Precast concrete facing, the precast concrete panel was used as wall facing with 1.7 m width and 1.5 m height. The precast concrete facing was modeled as plates element. The axial and bending stiffness are tabulated in Table 6.3.

Hexagonal wire mesh, hexagonal wire mesh reinforcement is referred to as flexible sheet material which is capable of resisting tensile stresses in the application of soil reinforcement. In the PLAXIS, this type of material is modeled as "geogrids". This element needs only an axial stiffness, AE, where A is the cross-sectional area per unit width in out of plane and E is the modulus of elasticity of the material. The axial stiffness was obtained from tensile test of hexagonal wire mesh in air, which was performed by [12]. However, due to the fact that the reinforcement, in this case,

Table 6.1 Selected parameters for soil in 2D FEM analysis

Depth (m)	Soil type	Model	Analysis	γ_{wet} (kN/m^3)	c'(kPa)	ϕ'	λ^a	κ^a	ν'	E'	Kh (10^{-4}) (m/d)	Kv (10^{-4}) (m/d)	OCR
0.0–1.5	Back fill	SSM	Undrained	17.6	14	23	0.14	0.015			25	12.5	4
1.5–2.5	Weather clay	MCM	Undrained	16.87	1	23			0.25	60,000	17	8.5	
2.5–3.5	Soft clay	SSM	Undrained	14.88	1	23	0.14	0.015			15.2	7.6	1.5
3.5–5.0	Soft clay	SSM	Undrained	14.88	1	23	0.15	0.015			15.2	7.6	1.4
5.0–7.0	Soft clay	SSM	Undrained	14.88	1	23	0.15	0.015			15.2	7.6	1.5
7.0–9.0	Soft clay	SSM	Undrained	14.88	1	23	0.14	0.015			15.2	7.6	1.8
9.0–11.0	Medium clay	MCM	Undrained	17.41	30	26			0.25	90,000	60	30	
11.0–20.0	Stiff clay	MCM	Undrained	19.84	30	26			0.25	12,000	0.8	0.4	
Emabnkment		MCM	Drained	17.1	10	20			0.33	10,000			
DMM pile		MCM	Undrained	18	121	36			0.33	50, 000	152	76	

Note *MCM* linear elastic perfectly plastic Mohr–Coulomb model
SSM Soft Soil Model
([a]): the permeability coefficients of sub-soil layers were taken as 10 times of the laboratory values

Table 6.2 Selected parameters for soil in 3D FEM analysis

Depth (m)	Soil type	Model	Analysis	γ_{wet} (kN/m^3)	c'(kPa)	ϕ'	λ^a	κ^a	ν'	E'	Kh (10^{-4}) (m/d)	Kv (10^{-4}) (m/d)	OCR
0.0–1.5	Back fill	SSM	Undrained	17.6	2	23	0.2	0.0400			25	12.5	1.5
1.5–2.5	Weather clay	MCM	Undrained	16.87	5	23			0.25	2500	17	8.5	
2.5–9.0	Soft clay	SSM	Undrained	14.88	1	23	0.2	0.0400			8	4	1.25
9.0–11	Medium clay	MCM	Undrained	17.41	8	23			0.25	5000	60	30	
11.0–20.0	Stiff clay	MCM	Undrained	19.84	30	26			0.25	9000	0.8	0.4	
Emabnkment		MCM	Drained	17.1	10	30			0.33	10,000			
DMM pile		MCM	Undrained	18	200	30			0.33	50,000	152	76	

Table 6.3 Reinforcement and concrete facing panel

Material type	EA (kN/m)	EI (kN m^2/m)
Hexagonal wire mesh	476.67	–
Concrete facing panel	4.35×10^6	8156.25

was not installed as continuous sheets along the length of the embankment, this EA value was reduced 1.5 times in the FEM simulation as the consequence. The axial stiffness of hexagonal wire applied in this analysis is tabulated in Table 6.3.

6.3.4 Soil/Hexagonal Wire Mesh Interface Models

Interface strength (R_{inter}), interaction of soil/hexagonal wire mesh in these modeling used elastic–plastic model to describe the behavior of interfaces. The Coulomb criterion was used to distinguish between elastic behavior, where small displacement can occur within the interface and plastic interface behavior when permanent slip may occur. The strength properties of interfaces are linked to the strength properties of a soil layer. Each data set has an associated strength reduction factor for interfaces (R_{inter}). The interface properties were calculated from the soil properties in the associated data.

Based on the full-scale pull-out test result of this study, the back-calculated R-value had a similar value of 0.67. This value was applied in the simulation of this reinforced embankment.

6.4 Degree of Consolidation

The average degree of consolidation of the soil–cement pile improved ground was obtained using the modified time factors derived from the analytical model of [7] that be substituted to the standard solution (or chart) or to any approximate solutions (e.g., [10]) of one-dimensional consolidation equation.

The actual load transfer mechanism must fall within the two extreme of either equal strain or equal stress conditions. Thus, the actual average degree of consolidation can be estimated by applying appropriate weighting factor to each average degree of consolidation from the two extreme conditions, namely equal stress and equal strain conditions. Consequently, the actual average degree of consolidation of the improved ground, $\overline{U}$, will be predicted using the following relationship:

$$\overline{U} = \alpha_\varepsilon(\overline{U}_{v,\varepsilon}) + \alpha_\sigma(\overline{U}_{v,\sigma}) \tag{6.1}$$

where α_ε and α_σ are the weighting factors of the average degree of consolidation corresponding to equal strain and equal stress conditions, respectively; $\overline{U}_{v,\varepsilon}$ is the average degree of consolidation under equal strain conditions calculated using the standard solution (or chart) or to any approximate solutions (e.g., [10] of one-dimensional consolidation equation with the time factor, $T_{v\varepsilon}$; and $(\overline{U}_{v,\sigma})$ is the average degree of consolidation under equal stress condition calculated using the time factor, $T_{v\sigma}$. Obviously, the sum of these two weighting factors, α_ε and α_σ, must be equal to unity.

The consolidation parameters, and the strength parameters of soil–cement piles used in the back-analyses, were derived from separate test piles, which were installed adjacent to the embankment. Pechgate et al. (2004) reported the following properties of the tested soil–cement piles: water content = 160%, γ_{wet} = 12.8 kN/m^3; q_u = 300–700 kPa and E_{50} = 60,000–120,000 kPa. From laboratory test, the specific gravity of the cement-admixed clay comprising the soil cement pile was found to be 2.65. Accordingly, the after-curing void ratio of cement-admixed clay materials for the soil–cement pile was obtained as 4.3, which was almost twice the void ratio of the natural clay. At this magnitude of after-curing void ratio of soil–cement piles, the coefficient of vertical permeability, $K_{v,p}$, ranged from 150 to 200 × 10^{-10} m/s, and the corresponding coefficient of consolidation, $C_{v,p}$, ranged from 200 to 400 m^2/yr [7]. In addition, for the surrounding clay, the coefficient of vertical permeability, $K_{v,c}$, ranged from 3 to 6 × 10^{-10} m/s, and the corresponding coefficient of consolidation, $C_{v,c}$, ranged from 1 to 3 m^2/yr.

6.5 Results and Discussion

6.5.1 Surface Settlements

At 397 days after the end of construction, the average surface settlements as measured from all the settlement plates on ground and on piles are 327.54 mm and 277.76 mm, respectively; while the average subsurface settlement at depths of 3.0 m and 6.0 m is 243.97 mm and 71.44 mm, respectively.

The observed settlement data were compared with analytical and numerical prediction methods. The analytical prediction was done using the method of [7] for the consolidation analysis of deep mixing improved ground, while the numerical method was done by FEM simulation.

The measured and the analytical prediction of surface settlements for S2 (on pile) and S6 (on clay) are plotted for comparison in Fig. 6.3. Significantly, the predicted surface settlements on ground and on pile using the analytical procedure of [7] are in good agreement with the observed data.

The observed and computed surface settlements on pile (at settlement plate S1) of the test embankment are illustrated in Fig. 6.4. Elastic settlements were observed during the early part of the construction period. The rate of settlement was higher

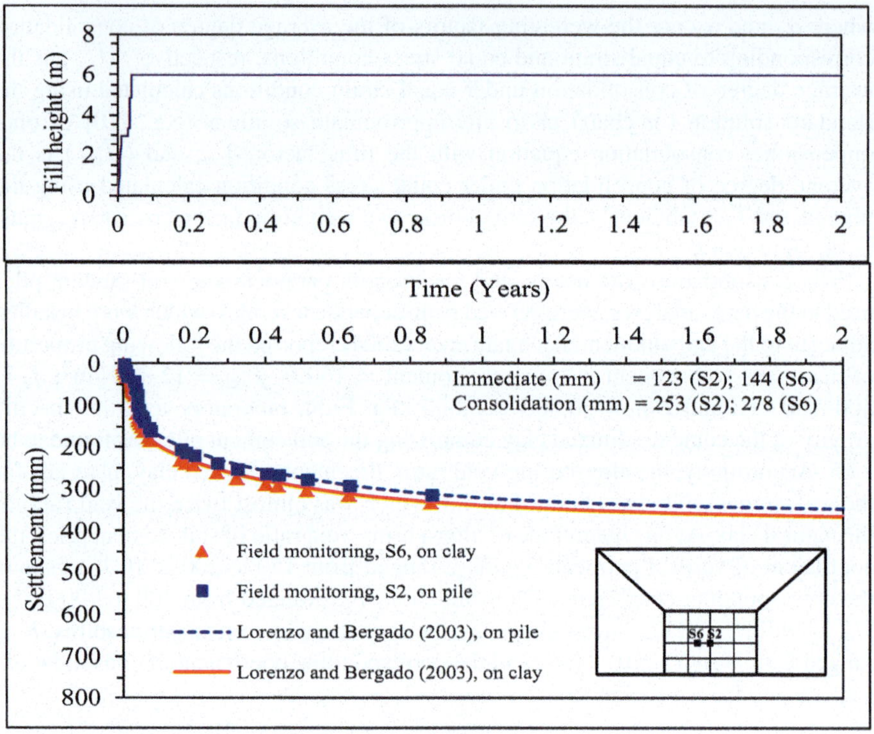

Fig. 6.3 Observed versus predicted surface settlements at S2 and S6

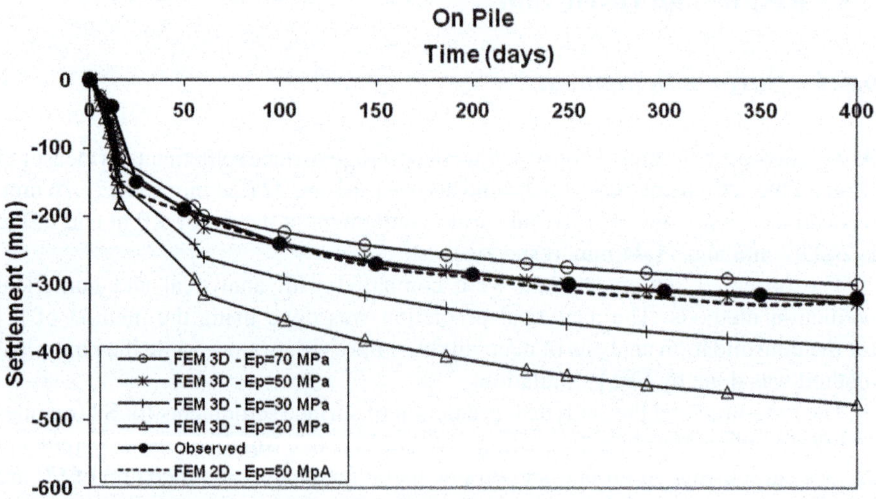

Fig. 6.4 Observed and computed surface settlement on pile with varying clay-cement pile modulus

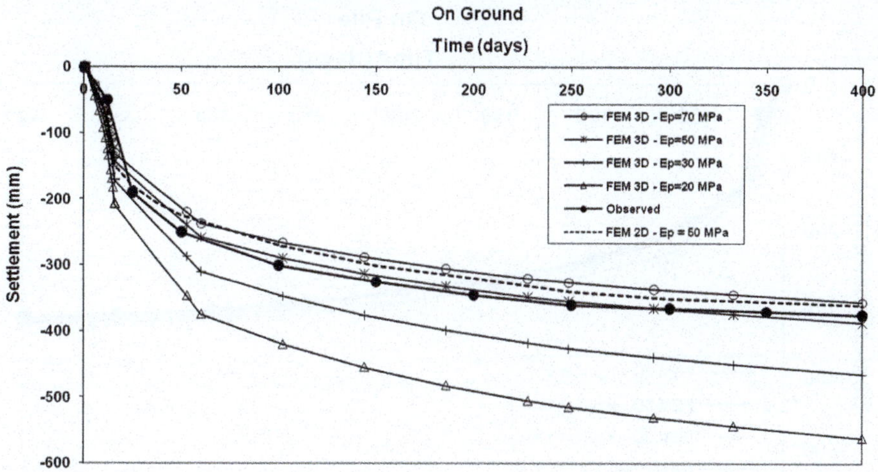

Fig. 6.5 Observed and computed surface settlement on ground with varying clay-cement pile modulus

at the beginning stage and slowed down progressively after 25 days. Due to the fact that stiffness of the foundation is much improved by clay-cement piles, its modulus plays an important role in the computation of vertical settlement on piles under embankment loading. The modulus of clay-cement pile was varied at 70, 50, 30, and 20 MPa to study the settlements' behavior. The computed settlement curve was in agreement with field observed settlement curve with time by adopting clay-cement pile modulus of 50 MPa. The performance of sensitivity analyses of settlement on pile showed that with higher stiffness of clay-cement pile, the lesser settlement has occurred. The overall settlement behavior/trend over time was identical but different in magnitude for different clay-cement pile modulus.

The observed and computed surface settlements on ground (clay surrounding the clay-cement piles) are illustrated in Fig. 6.5. The 3D computation results showed that computed settlements agreed well with observed settlements with clay-cement pile modulus of 50 kPa. The influence of clay-cement pile modulus on surface settlement on ground was studied by varying its modulus at 70, 50, 30, and 20 MPa. As expected, the results showed that settlements on ground were reduced with higher clay-cement pile modulus and maintained the same settlement trends over time.

According to previous study (Uddin 1995), the permeability of cement-admixed clay can be higher than that of the base clay and thus, the rate of consolidation is increased for the DMM improved ground. Sensitivity study of the effect of clay-cement pile permeability was carried out with ratio, of $k_{\text{pile}}/k_{\text{soil}}$ at 5, 18.8, and 100. For both surface settlements on pile and on ground, the computed settlements agreed well with the observed data at $k_{\text{pile}}/k_{\text{soil}}$ equal to 18.8 as shown in Figs. 6.6 and 6.7. The effect of permeability of clay-cement pile to surface settlements on pile and on ground was not significant comparing with clay-cement pile modulus.

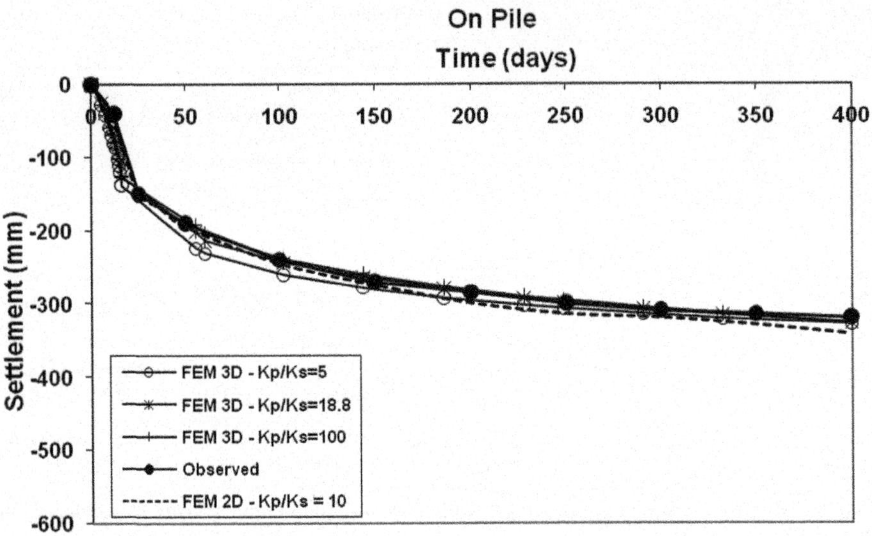

Fig. 6.6 Observed and computed surface settlement on pile with varying clay-cement pile permeability

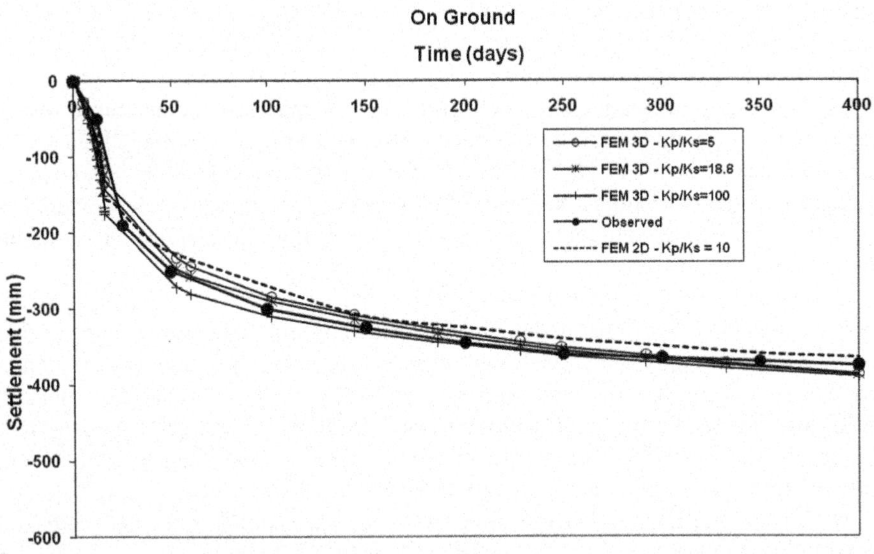

Fig. 6.7 Observed and computed surface settlement on ground with varying clay-cement pile permeability

As demonstrated in the aforementioned cases, the increment of settlement was more apparent when clay-cement pile modulus was reduced by 20 MPa (decreased from 50 to 30 MPa) compared with when clay-cement pile modulus was increased by 20 kPa (increased from 50 to 70 MPa). This phenomenon showed that maintaining a good quality of mixing during DMM pile installation is essential.

The computed surface settlement on pile in 2D FE analyses was in agreement with 3D FE analyses by adopting clay-cement pile modulus of 50 MPa, as shown in Figs. 6.4 and 6.5. For surface settlement on ground, the computed settlement in 2D FE analyses was in agreement with 3D FE analyses by adopting clay-cement pile permeability, k_{pile}/k_{soil} at value 10, and k_{pile}/k_{soil} at value 18.8; respectively, as shown in Figs. 6.6 and 6.7. Thus the clay-cement pile has higher permeability in 3D conditions.

6.5.2 Subsurface Settlements

The observed and computed subsurface settlements at 3.0 m are illustrated in Figs. 6.8 and 6.9 with varying parameters of clay-cement pile modulus, E_p at 70 Mpa, 50 Mpa, 30 Mpa, 20 Mpa and permeability, k_{pile}/k_{soil} at 5, 18.8 and 100, respectively. In both cases, the computed settlements were in agreement with observed data at clay-cement pile modulus, $E_p = 50$ Mpa, and permeability, $k_{pile}/k_{soil} = 18.8$, respectively, in 3D FE analyses. As expected, lower subsurface settlements at 3.0 m occurred with higher clay-cement pile modulus as shown in Fig. 6.7. Figure 6.8 shows that influence of

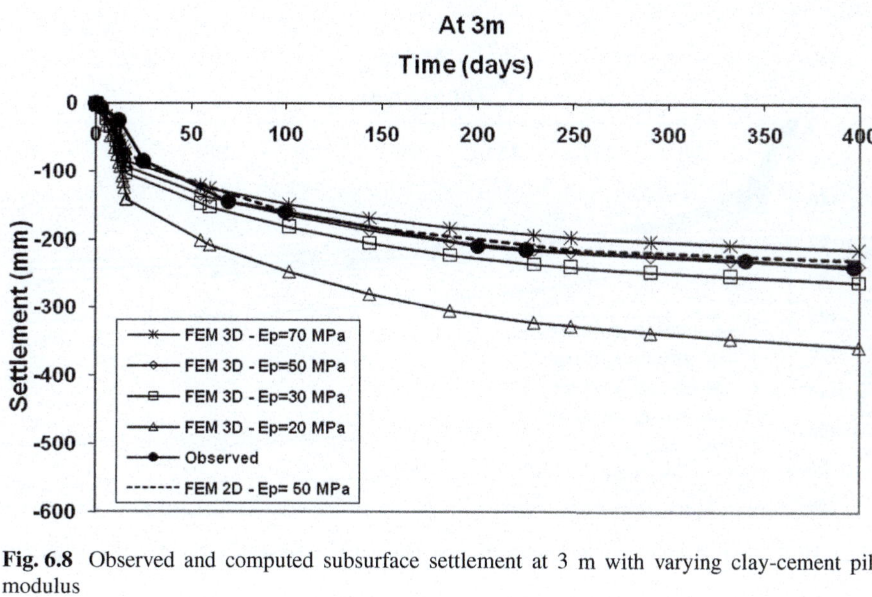

Fig. 6.8 Observed and computed subsurface settlement at 3 m with varying clay-cement pile modulus

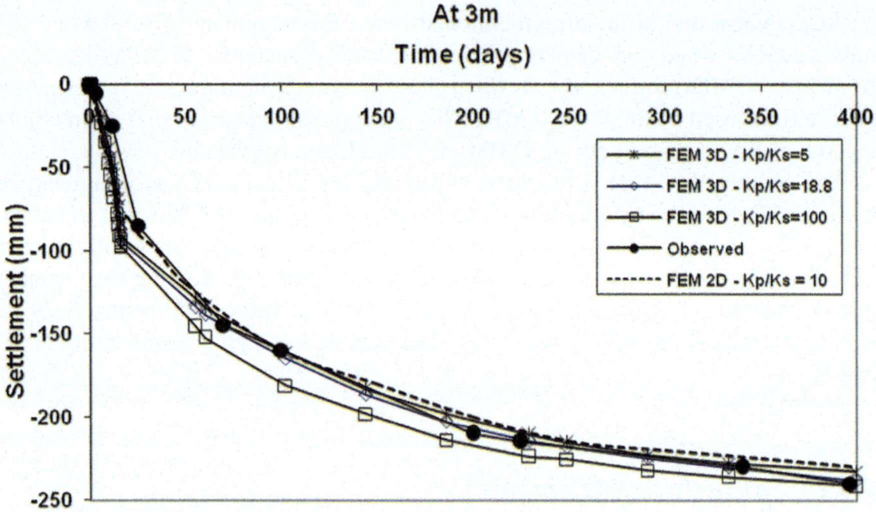

Fig. 6.9 Observed and computed subsurface settlement at 3 m with varying clay-cement pile permeability

clay-cement pile modulus to subsurface settlement at 3.0 m is more significant than clay-cement pile permeability.

The observed and computed subsurface settlements at 6.0 m are illustrated in Figs. 6.10 and 6.11 with varying parameters of clay-cement pile modulus, E_p at 70 Mpa, 50 Mpa, 30 Mpa, 20 Mpa and permeability, k_{pile}/k_{soil} at 5, 18.8 and 100,

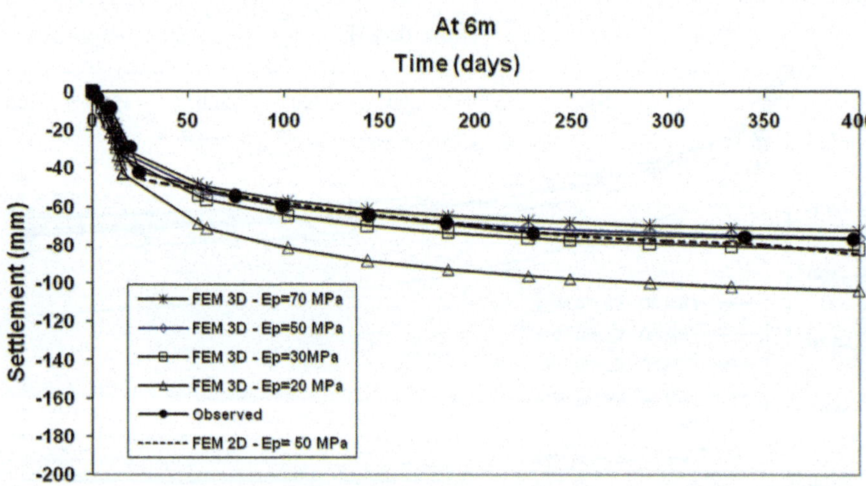

Fig. 6.10 Observed and computed subsurface settlement at 6 m with varying clay-cement pile modulus

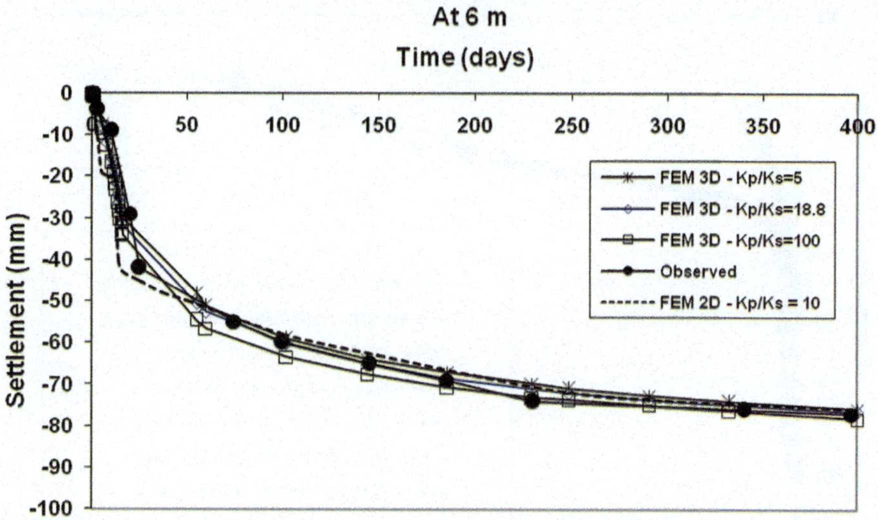

Fig. 6.11 Observed and computed subsurface settlement at 6 m with varying clay-cement pile permeability

respectively. In both cases, the clay-cement pile modulus, $E_p = 50$ Mpa, and permeability, $k_{pile}/k_{soil} = 18.8$ agreed well with the observed data under 3D conditions. As expected, lower subsurface settlement at 6.0 m occurred at higher clay-cement pile modulus as shown in Fig. 6.10. Under 3D conditions, the influence of clay-cement pile modulus to subsurface settlement is more significant at 3.0 m than that at 6.0 m. In both cases, the influence of clay-cement pile modulus to subsurface settlement at 3.0 m is more significant than clay-cement pile permeability. For subsurface settlement at 6.0 m, the influence of clay-cement pile permeability to settlement is insignificant as shown in Fig. 6.11.

The computed subsurface settlement at 3 and 6 m in 2D FE analyses were in agreement with 3D FE analyses by adopting clay-cement pile modulus as shown in Figs. 6.8 and 6.10. For subsurface settlement at 3 and 6 m, the computed settlement in 2D FE analyses was in agreement with 3D FE analyses by adopting clay-cement pile permeability, k_{pile}/k_{soil} equal to 10 and k_{pile}/k_{soil} equal to 18.8; in 2D and 3D, respectively, as shown in Figs. 6.9 and 6.11. As expected, the clay-cement pile has higher permeability in 3D conditions.

6.5.3 *Lateral Movement*

The observed and computed lateral movement of test embankment, as well as subsoil foundation, is illustrated in Fig. 6.12 at about 400 days after construction. As expected, the coefficient of interaction of soil reinforcement influenced the lateral

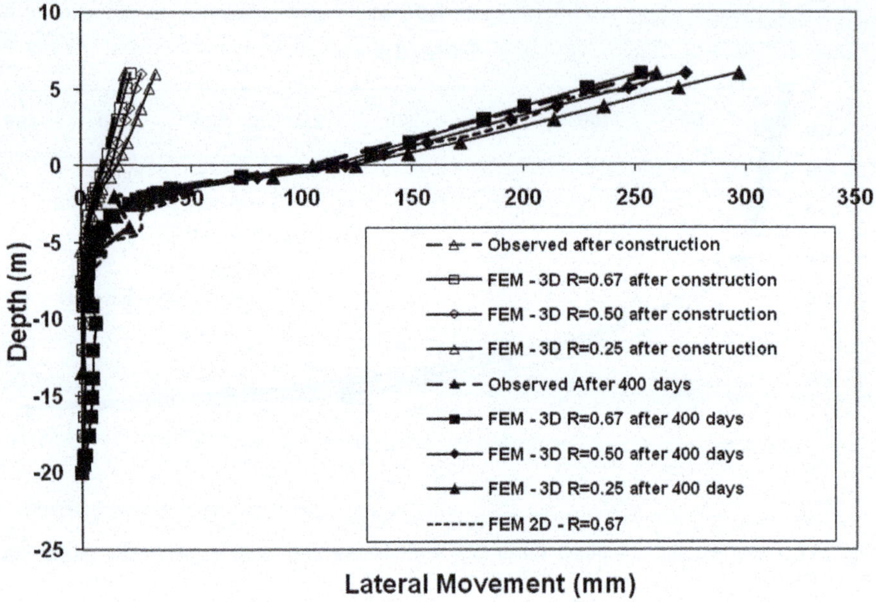

Fig. 6.12 Observed and computed lateral movement with varying reduction factor, *R*

displacement behavior of the reinforced embankment. Sensitivity study was carried out to study the influence of the reduction factor, *R* at values 0.25, 0.50, and 0.67 at time after construction and 400 days after construction, respectively. In both cases, the computed lateral movements in 3D FE analyses generally well agreed with field observations at soil/reinforcement interaction coefficient of 0.67. The lateral movement is influenced by the coefficient of interaction between the soil and hexagonal wire mesh reinforcement. Higher interaction coefficient mobilized greater pullout and direct shear strength at the interfaces and thus, reduced the lateral movement due to the fact that the whole reinforced soil mass acted more like a rigid body.

The computed lateral movement in 2D FE analyses with interaction coefficient, *R*, of 0.67 was in agreement with computed lateral movement in 3D FE analyses with similar interaction coefficients. However, the computed lateral movement in 2D FE analyses overestimated the computed value in 3D FE analyses at 1.5–2.5 m depth in the foundation soil as shown in Fig. 6.12.

6.5.4 Excess Pore Pressure

The observed and computed excess pore pressure in 3D FE analyses at 3 and 6 m are illustrated in Figs. 6.13 and 6.14. In general, all simulated results in 3D analyses overestimated the excess pore pressure observed except at 3.0 m depth whereby

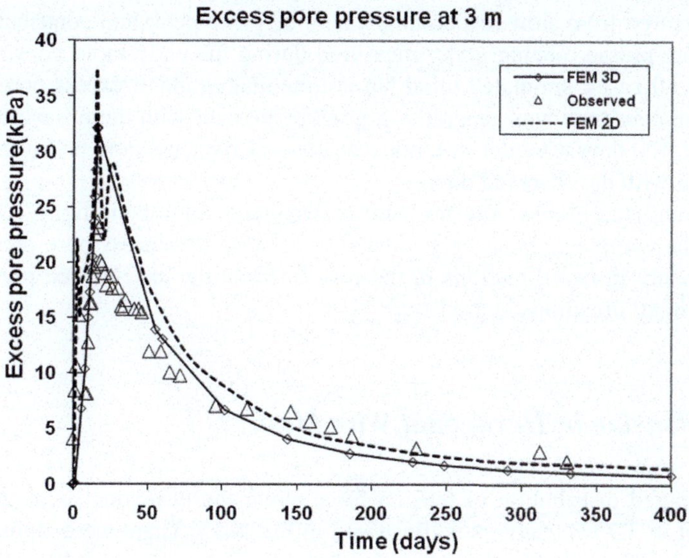

Fig. 6.13 Observed and computed excess pore pressure at 3 m

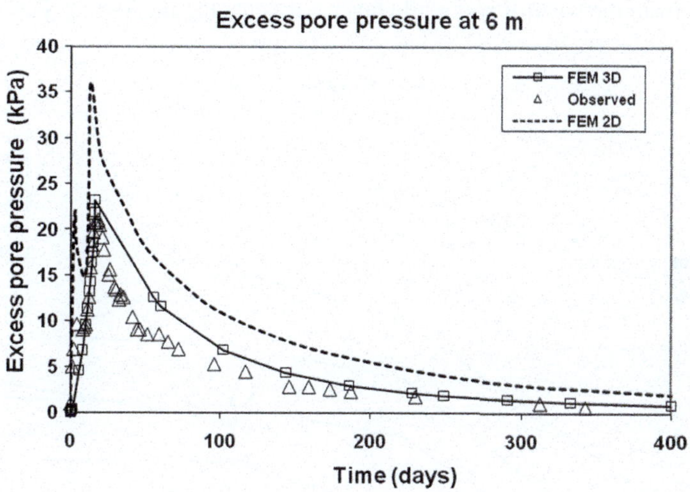

Fig. 6.14 Observed and computed excess pore pressure at 6 m

the excess pore pressure was underestimated in period after 100 days onwards. The computed excess pore pressures in early period during construction at 3 and 6 m were overestimated than that of the observed data. This can be explained by the undrained analysis which was adopted in the simulation during the staged construction period in PLAXIS 3D Tunnel Program. In reality, the excess pore pressure in the foundation soil started to dissipate during construction. In addition, the excess pore pressures were measured from slow responding standpipe piezometer and contributed to the excess pore pressure being under measured during the early rapid construction in stages. In all cases, simulated trend but not the magnitude of excess pore pressure dissipation over time was generally in good agreement with the observed data. At the end of 400 days after construction, the computed excess pore pressures were in agreement with the observed data.

The computed excess pore pressure under plane strain condition (2D) overestimated the computed value in 3D analyses because the excess pore pressure can dissipate only in two directions in the former while the excess pore pressure can dissipate in all directions in the latter.

6.5.5 Tension in Hexagonal Wire Mesh

The computed distribution of tension force along the hexagonal wire mesh reinforcement in 3D FE analyses is illustrated in Fig. 6.15. There were eight layers of reinforcement installed in the test embankment, namely, layers 1–8 where layer 1 being at the lowest elevation at 0.375 m and layer 8 being at the highest elevation at 5.625 m from ground. Consequently, layer 1 was having maximum overburden pressures from the backfill soil while layer 8 was having the lowest. The simulated

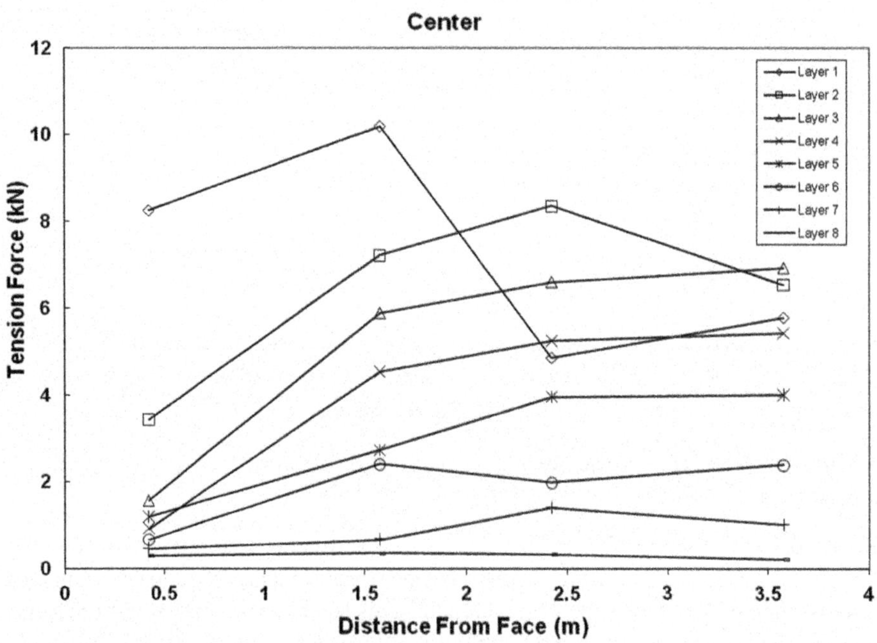

Fig. 6.15 Simulated tension force along reinforcement

results show that the tension force was maximum at layer 1 and minimum at layer 8 as expected. Figure 6.16 shows that observed data was more or less in agreement with simulated results at layers 1, 3, 5, and 7 with some signs of overestimation at locations near the wall face and underestimation at locations away from the wall face. This is due to the limitations of the simulation during the consolidation process. The maximum tension line generally agreed well with coherent gravity bilinear failure plane.

6.6 Conclusions

The case history of the full-scale reinforced embankment on jet grouting cement mixing piles improved soft Bangkok clay with full instrumentations had been considered and its behavior has been monitored for about 400 days after construction. The following conclusions were derived;

(1) The performance of the test embankment revealed that the cement mixing pile foundation has effectively reduced the settlement of reinforced soil test embankment by 70%. The local differential settlement between cement mixing piles and adjacent surrounding soil was 25–60 mm, which was about 8–20% of the average settlement.

(2) Numerical simulation in 2D FEM condition revealed that the values of cement-clay pile modulus, *E*, its permeability ratio, $k_{pile}/k_{soil,}$ and interaction coefficient of hexagonal wire mesh with backfill soil, *R*, are 50 Mpa, 10, and 0.67, respectively, agreed well with the observed data.

(3) Numerical simulations in 3D FEM condition, k_{pile}/k_{soil}, and interaction coefficient of hexagonal wire mesh reinforcement with backfill soil, *R*, are 50 MPa, 18.8, and 0.67, respectively, agreed well with observed data.

(4) The predicted surface settlements were sensitive to the clay-cement pile modulus more that the permeability. Both excess pore water pressure predictions in 2D and 3D conditions were over predicted mainly because undrained analysis was adopted in the simulation during the early staged construction. In reality, the excess pore pressure in the foundation soil started to dissipate during the stage of construction. In addition to that, slow measurement of excess pore pressure from slow responding standpipe piezometer also contributed to the under measured of excess pore pressures during rapid staged construction.

(5) In 3D FE simulations, the predicted tension forces along the hexagonal wire reinforcement were slightly greater than the observed field data due to the limitations of the simulation during the consolidation process. The maximum tension line generally agreed well with coherent gravity bilinear failure plane.

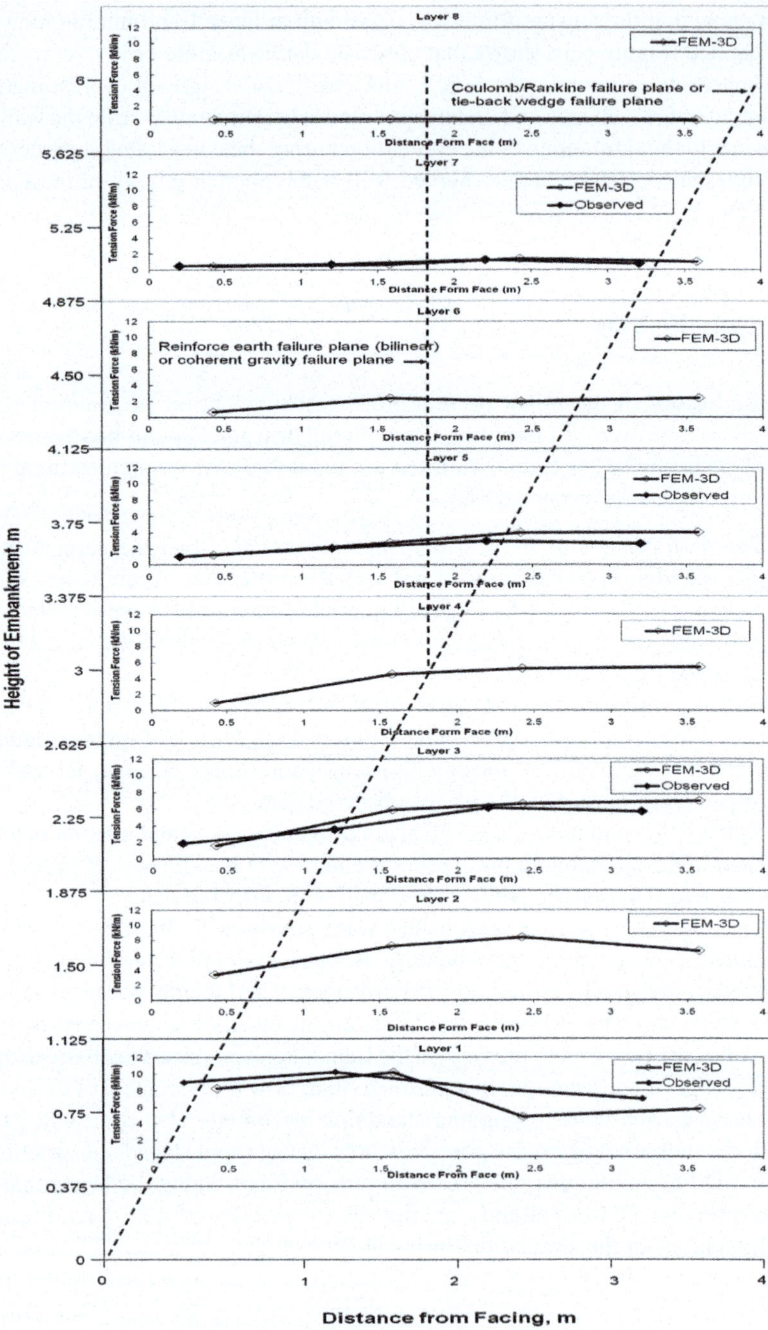

Fig. 6.16 Observed and computed maximum tension force in reinforcement

References

1. Bergado, D.T., Lorenzo, G.A., Lai, Y.P., Pun A.: Deep mixing method using cement admixture at higher water content as foundation of hexagonal wire reinforced embankment. In: Proceedings of the 2nd Advances in Soft Soil Engineering and Technology, Malaysia, July 2003, pp. 317–334 (2003a)
2. Bergado, D.T., Lorenzo, G.A., Balasubramaniam, A.S.: Behavior of cement and lime treated soft Bangkok clay. In: Proceedings of the 12th Asian Regional Conference on Soil Mechanics and Geotechnical Engineering, Singapore, August 2003, Vol. 1, pp. 449–452 (2003b)
3. Broms, B.B.: Stabilization of soft clay with lime columns. In: Proceedings of the Seminar on Soil Improvement and Construction Techniques in Soft Ground, Nanyang Technological Institute, Singapore (1984)
4. Kamon, M., Bergado, D.T.: Ground improvement techniques. In: Proceedings of the 9th Asian Regional Conference on Soil Mechanics and Foundation Engineering, Bangkok, Thailand, Vol. 2, pp. 526–546 (1991)
5. Lorenzo, G.A., Bergado, D.T.: New consolidation equation for soil-cement pile improved ground. Can. Geotech. J. **40**(2), 265–275 (2003)
6. Pechgate, K., Modmoltin, C., Bergado, D.T., and Voottipruex, P.: Effect of Jet Grouting on Adjacent Soils of Soil Cement Column. In: Proceeding of the International Symposium on Engineering Practice and Performance of Soft Deposits (IS Osaka 2004), Osaka, Japan, pp. 247–250 (2004)
7. Porbaha, A.: State of the art in deep mixing technology. Part I: Basic concepts and overview. Ground Improv. **2**, 81–92 (1998)
8. Porbaha, A.: State of the art in deep mixing technology. Part II: Applications. Ground Improv. **2**, 81–92 (1998)
9. Sivaram, B., Swamee, P.: A computational method for consolidation coefficient. Soil Found. **17**(12), 48–52 (1977). Tokyo, Japan
10. Tong, B. Schaefer, V. Liu, Y., Han, B.: Optimization of deep mixing design for seismic liquefaction mitigation of Caisson walls. Geomatics Nat. Hazards Risk 287–313 (2019)
11. Uddin, M. K. (1995). Strength and deformation characteristics of cement-treated Bangkok clay. D. Eng. Dissertation No. GT-94-1, Asian Institute of Technology, Bangkok, Thailand
12. Wongsawanon, T.: Interaction between Hexagonal Wire Mesh Reinforcement and Silty-Sand Backfill. M.Eng. Thesis No. GE-97-14, Asian Institute of Technology, Thailand (1998)
13. Yamashita, K., Shigeno, Y., Hamada, J.: Effect of improved area ratio of deep mixing wall grid on pile stress in piled raft system under strong earthquakes. Takenaka Technical Research Report, No. 74 (2018)

Chapter 7
Reinforcement of Runway Embankment by Using Geosynthetics in Sulawesi, Indonesia

Tjie-Liong Gouw

7.1 Introduction

Indonesia is an archipelago country comprising 17,504 islands. This condition makes sea and air transportation a crucial factor in its national development. The fact that Indonesia is located right in a high seismicity region (Fig. 7.1) with many geotechnically difficult ground conditions such as peats, soft clays (Fig. 7.2), liquefiable sandy ground, clay shale formation, etc., the rapid developments throw great challenges to the local geotechnical engineers. One of those geotechnical challenges is to build airports where the runway must be built on hilly terrains requiring massive cut and fill operations. Very often 20 m to over 40 m high man-made slopes had to be built in an earthquake-prone area and on a difficult sub-soil conditions. Following the experiences in Japan where it is reported that geosynthetic reinforced soil slopes can accommodate larger deformation and perform better under seismic loading than conventional rigid retaining walls [9–11, 16], this reinforcing system is often adopted for stabilizing the high man-made slopes in Indonesia. A case study on the application of geosynthetic reinforced slopes of up to 37 m high on difficult clay shale formation is presented and discussed here.

T.-L. Gouw (✉)
Universitas Katolik Parahyangan, Bandung, Indonesia
e-mail: gouw_tjie@unpar.ac.id

H. Hazarika et al. (eds.), *Practices and Trends in Ground Improvement Techniques*, Developments in Geotechnical Engineering,
https://doi.org/10.1007/978-981-19-3322-6_7

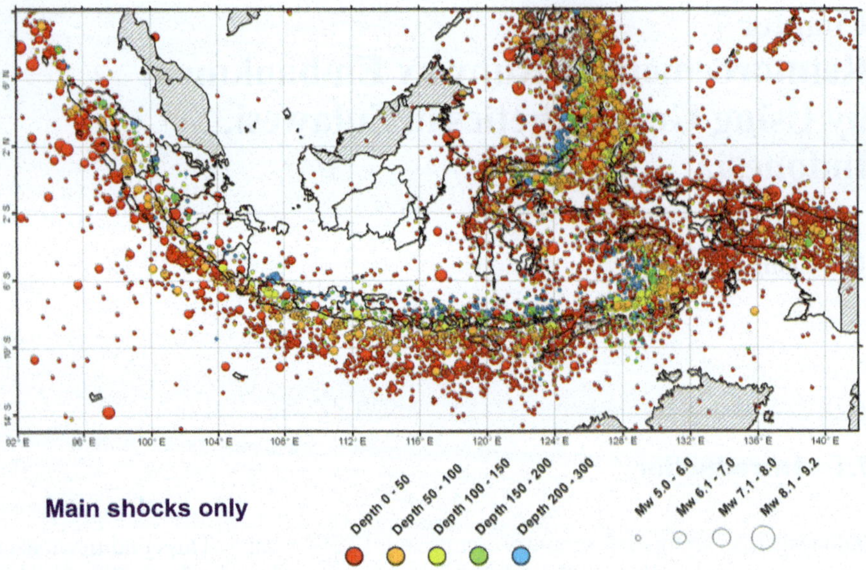

Fig. 7.1 Recorded seismic events in Indonesia [7]

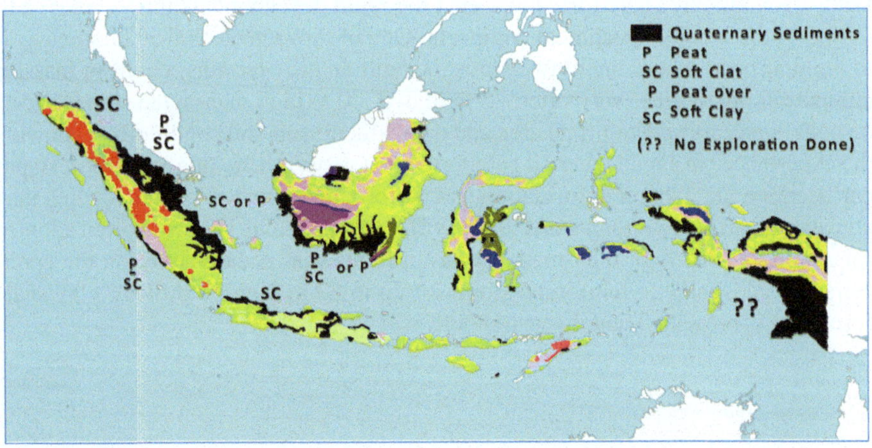

Fig. 7.2 Distribution of peat and soft clay in Indonesia (redrawn after [2])

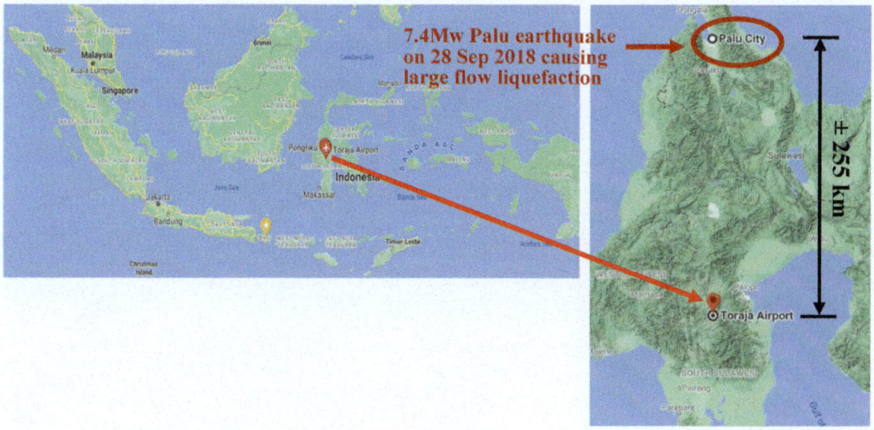

Fig. 7.3 New Toraja airport location

7.2 Building 37 m High Reinforced Soil Slopes Over Clay Shale Formation

7.2.1 Project Location

The New Tana Toraja airport is located in South Sulawesi province (Fig. 7.3). It is about 255 km straight line distance from Palu city where a 7.4 Mw earthquake occurred on 28 Sept 2018 causing large flow liquefaction. The purpose of building the new airport is to enhance the tourism industry of Tana Toraja with its breath-taking landscapes, ancestral houses, burial rituals, and unique traditions of the Toraja peoples. This new airport project, with its 210 m wide and 2 km long runway, was started in 2014 and completed in 2020. The first plane landing took place on August 12, 2020.

7.2.2 Geotechnical Problems

Owing to its hilly terrain, massive cut and fill operations and embankment slopes ranging from 15 to 37 m high had to be built to construct a flat runway (Fig. 7.4). The initial soil investigation results showed the existence of strong foundation soil at just 3 7 m depth with SPT blow counts > 60 blows/ft and CPT cone resistance > 15 MPa and it was reported as a "soft rock" formation (Fig. 7.5). Judging that the foundation soil was of a very strong layer, the first stage of 15 m high unreinforced slope was built. However, not long after, the slope collapsed in a base failure mode. The author was flown in to investigate what was the cause of the failure and to provide a solution. Upon investigation, it was found that what was supposed as soft rocks

Fig. 7.4 Overview of New Tana Toraja airport project area

were problematic clay shales that significantly lose their shear strength after just a few days of exposure to outside weather (Fig. 7.6).

Figure 7.7 shows the newly cut slope exposing the freshly exposed clay shale formation at the jobsite. Figure 7.8 shows another cut slope that had been exposed for a couple of weeks, rainwater eroded the slopes, and gullies were formed when the surface runoff flowed over the slopes. The exposed clay shales disintegrated and rapidly lost their shear strength. Figure 7.9 shows the freshly excavated pieces of rock-like clay shales, while Fig. 7.10 shows the disintegrated clay shales easily crumble off by just a light pressure of fingers. Gartung [3] reported fresh unexposed clay shale can have drained cohesion, c', and drained friction angle, ϕ', up to 85 kPa and 41°, respectively. However, when prolonged exposure to the atmosphere takes place, it weathers rapidly, and its shear strength can drop to zero cohesion and friction angle of merely 9° (Fig. 7.11). Reference [15] showed unsoaked clay shale can reach a peak cohesion of 5500 psf (260 kPa) with a 39° friction angle, while under soaked

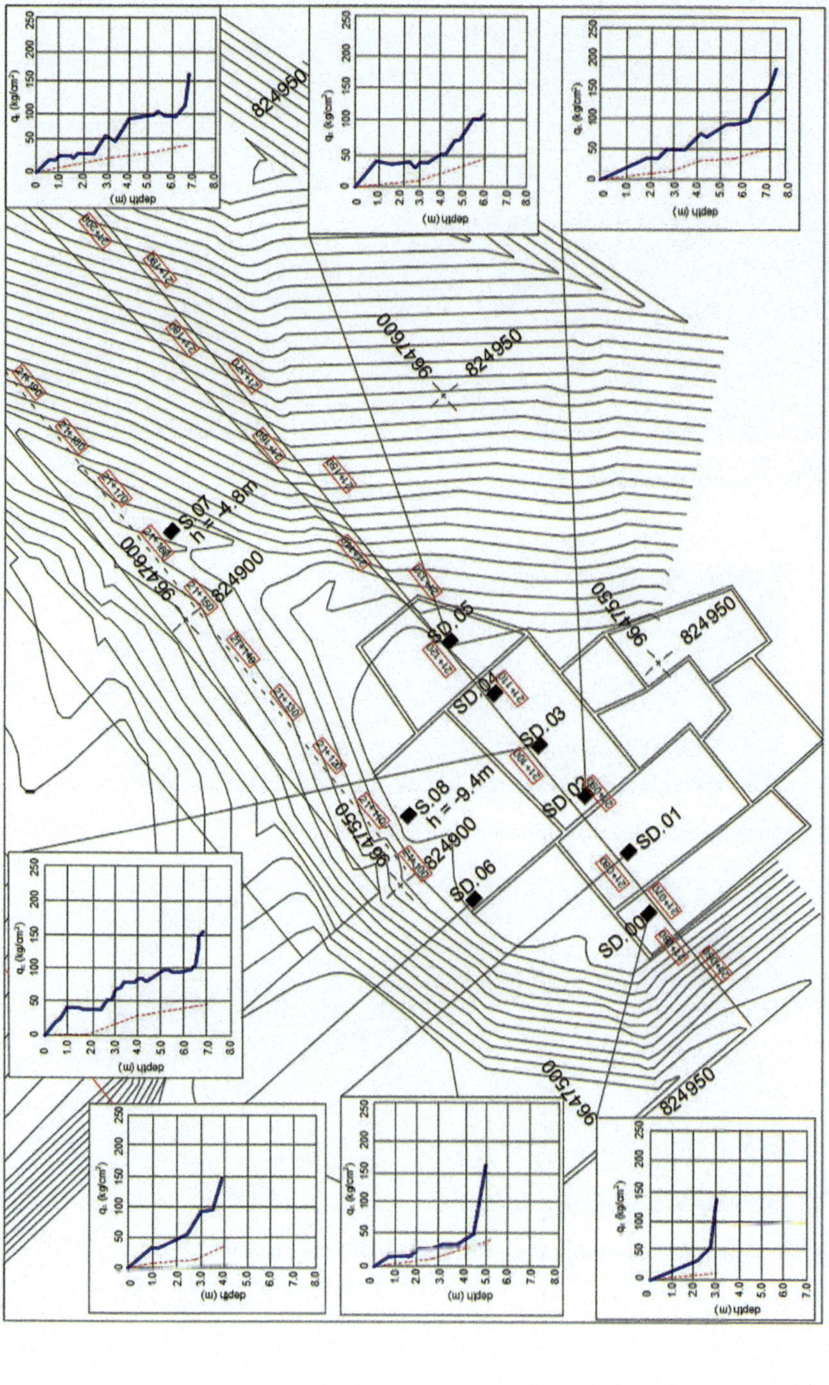

Fig. 7.5 Initial soil investigation points and CPT results

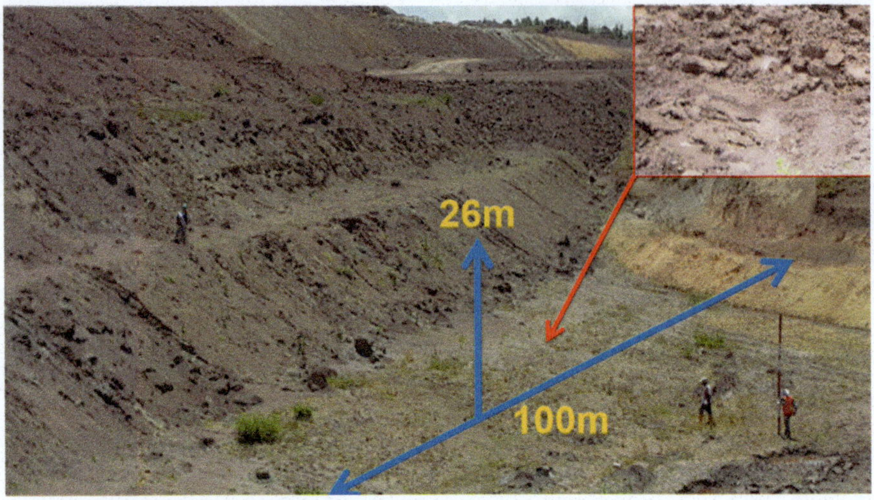

Fig. 7.6 Deteriorated foundation soil

Fig. 7.7 Freshly exposed clay shale formation in a newly cut slope

Fig. 7.8 Weathered clay shale formation a couple of weeks after exposure

Fig. 7.9 Pieces of fresh rock-like clay shale

Fig. 7.10 Weathered clay shale easily crumbles off by light pressure of fingers

conditions the cohesion disappears, and its residual friction angle drops to a low value of 15° (Fig. 7.12).

Another geotechnical problem is the seismicity of the project site. According to the Indonesian seismic code 13 the site is located in an area with a peak ground acceleration (PGA) of 0.2–0.25 g for a return period of 500 years as shown in Fig. 7.13.

Facing the fast strength degradation of the clay shale foundation soil, forming a stable unreinforced embankment slope requires a very gentle inclination of 1V: 5H or even gentler in a seismic condition. A costly land acquisition issue made this option unsuitable. Therefore, to build the high slopes, the technical advisory committee of the project gave the following design criteria:

- The slope shall be reinforced with retaining soil structures that can accommodate differential settlements, especially under earthquake conditions.
- A slightly higher PGA than the national standard, i.e., PGA of 0.30 g must be considered in the design. In pseudo-static analysis, a horizontal seismic coefficient of $k_h = 50\%$ PGA can be taken [17].

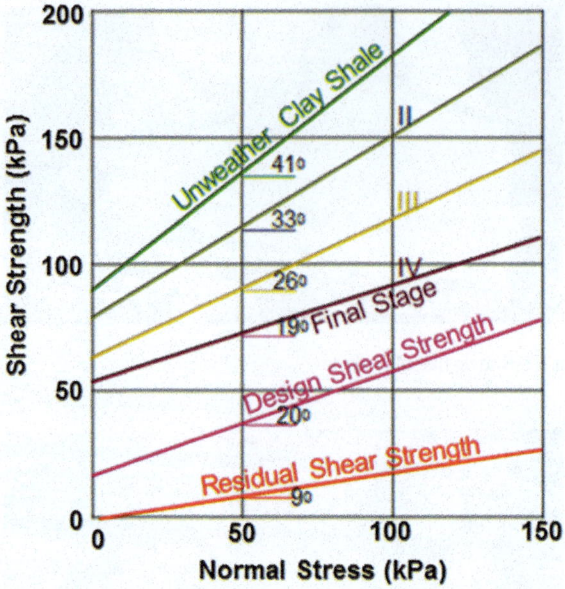

Fig. 7.11 Shear strength of fresh unweathered and weathered clay shale (redrawn after [3])

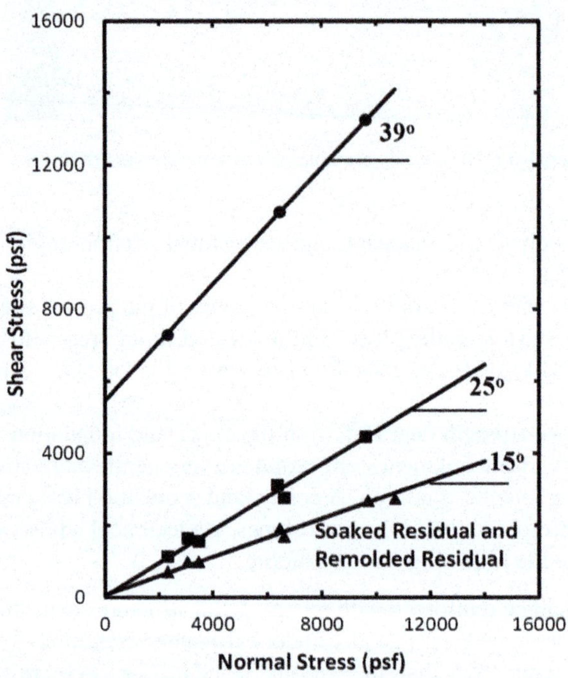

Fig. 7.12 Shear strength of soaked and unsoaked clay shale [15]

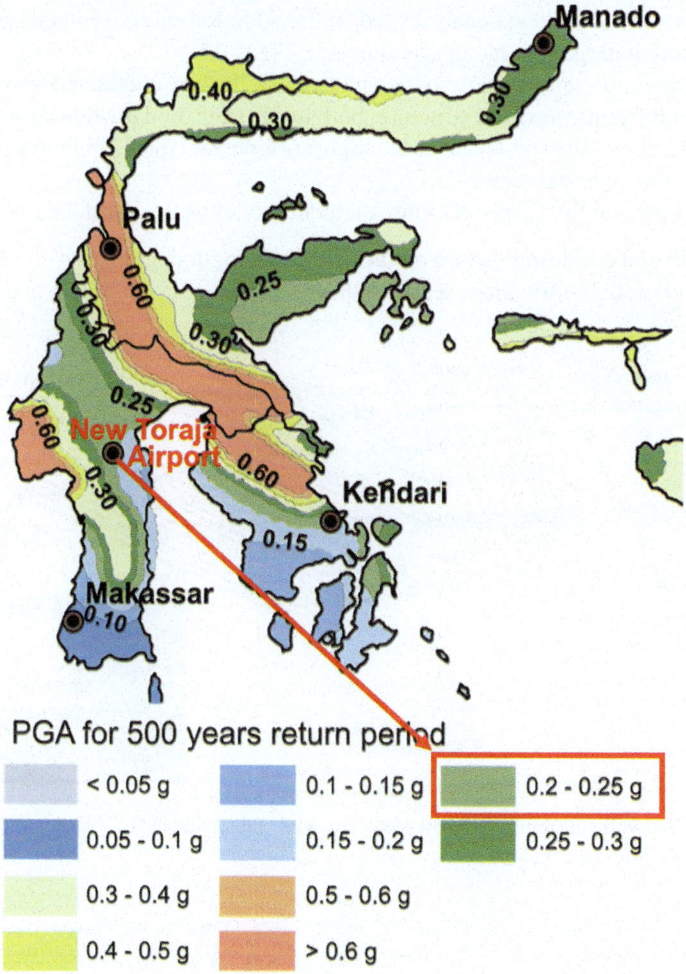

Fig. 7.13 Peak Ground Acceleration (PGA) at Tana Toraja Site [13]

- Locally available backfill material which is cohesive must be used.
- The construction of the high slopes must be cost-effective, be carried out in the shortest time possible, and maximize the employment of the local workers who are generally unskilled labor.

7.2.3 *Design of Geosynthetics Reinforced Slopes*

Three construction alternatives were considered, i.e., reinforced concrete buttress retaining wall, contiguous bored pile, and geosynthetics reinforced soil structure.

Detailed technical and cost analyses led to the adoption of geosynthetic reinforced soil structure with gabion facings as shown in Fig. 7.14.

Rigorous finite element analyses using PLAXIS 2D were conducted to evaluate the stability of the reinforced soil structure, both under static and earthquake conditions. Figure 7.15 shows the soil layers of the highest slope, i.e., the 37 m high slope. Table 7.1 shows the input parameters.

A few notes on the Plaxis 2D finite element modeling are given below:

- The sub-soil condition: Based on the known strength degradation characteristics of the clay shale formation, reduced shear strength of $c' = 20$ kPa and $\phi' = 17°$

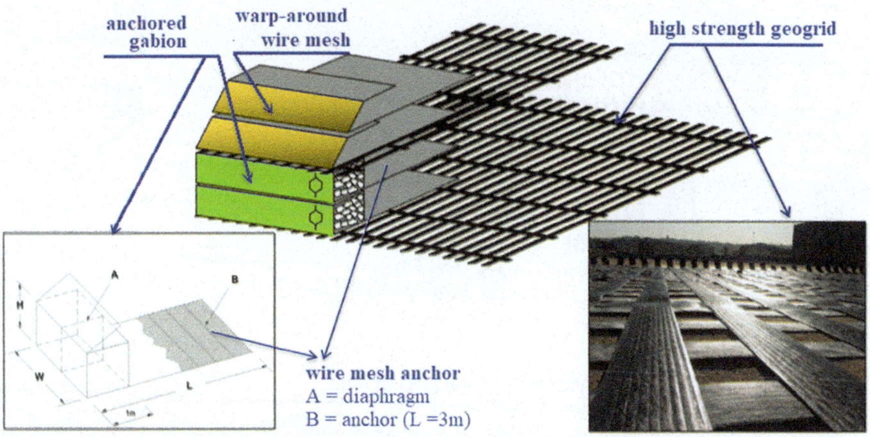

Fig. 7.14 Geosynthetic reinforced soil structure with gabion facings (courtesy of Maccaferri Indonesia)

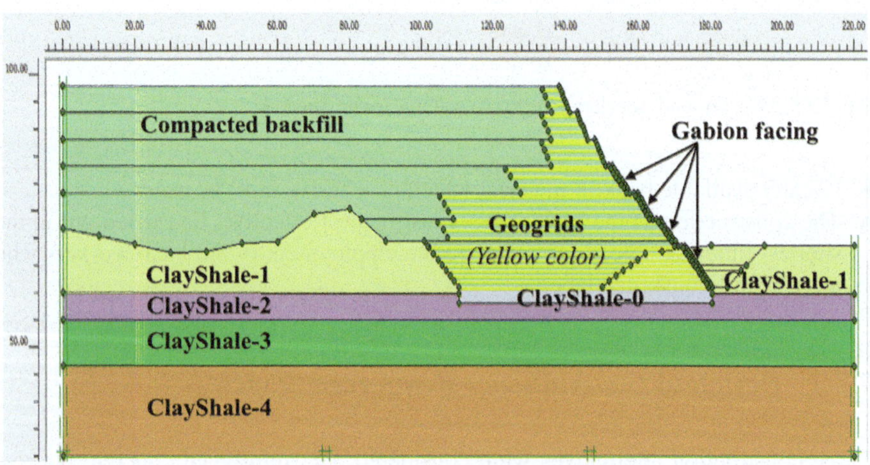

Fig. 7.15 Soil layers at the 37 m high slope

Table 7.1 Soil data input parameters [5]

Soil data input—Mohr–Coulomb soil model

ID	Name	Type	γ_{unsat}	γ_{sat}	v	E_{ref}	c'	ϕ'	ψ
			kN/m^3	kN/m^3	[–]	kN/m^2	kN/m^2	[°]	[°]
1	Compacted backfill	Drained	17	17	0.30	40,000	42	16	0
2	Clay shale-0	Drained	17	17	0.30	10,000	20	17	0
3	Clay shale-1	Drained	17	17	0.30	10,000	20	24	0
4	Clay shale-2	Drained	18	18	0.30	20,000	20	30	0
5	Clay shale-3	Drained	19	19	0.35	20,000	20	32	0
6	Clay shale-4	Drained	20	20	0.35	100,000	40	37	5

Gabion—Linear elastic model

ID	Name	Type	γ_{unsat}	γ_{sat}	v	E_{ref}	R_{inter}
			kN/m^3	kN/m^3	[–]	kN/m^2	[°]
7	Gabion filling	Drained	17.5	17.5	0.35	40,000	1

Geogrids

ID	Name	Type	EA	N_p
			kN/m	kN/m
1	Geogrid 300 kN/m	Plastic	3900	272

Note
γ_{unsat} = unsaturated unit weight
γ_{sat} = saturated unit weight
v = Poisson's ratio
E_{ref} = Young's modulus at a reference pressure of 100 kPa
c' = drained (effective) cohesion
ϕ' = drained (effective) friction angle
ψ = dilation angle
R_{inter} = interface coefficient
EA = axial (tension) stiffness of the geosynthetic
N_p = limiting tensile strength of the geosynthetic

was adopted for the first upper 3 m of the clay shale, i.e., the layer named as "clay shale-0" in Fig. 7.15. Below that, greater strengths of the clay shale layers were adopted.

- The facing elements: As gabions are used for the facing elements, it is improper if it is modeled as plate elements because it will result in bending moment be induced in the facings where in reality there will be no bending moment in the gabions. Therefore, each gabion block is modeled as a soil cluster with its actual dimension. The material model of the gabion is taken as linear elastic as the stones are placed inside a strongly tied wire-mesh box. Since water can flow through the gabion the material drainage type is chosen as drained. Interface element is added between two blocks of the gabions as the blocks may slide/slip one another. An interface factor of $R_{inter} = 0.9$ was introduced (see Eqs. 7.1 and 7.2 below for the definition of R_{inter}).

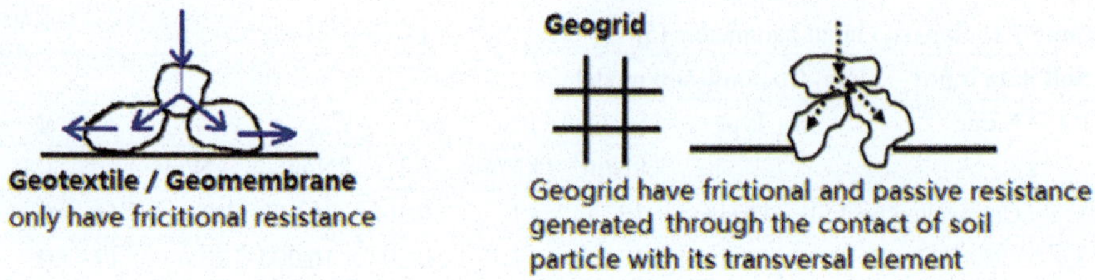

Fig. 7.16 Frictional resistance between soil and geosynthetics [6]

- The geogrids and interface elements: Geosynthetics work as tensile elements. In Plaxis the tensile element, for an unknown reason, is somehow named as "geogrids" element which actually can be used to model any material that withstands only tensile force. In this project, geogrids were used as the reinforcing element (see Fig. 7.14). The question then arose whether the interface element in Plaxis should be implemented or not. The interface element in Plaxis has two functions. The first function is to reduce contact friction between the backfill material (soil) and the structural material which in this case is the geogrids. The magnitude of the contact friction reduction is introduced by an interface coefficient, termed R_{inter} with a value of $0.01 < R_{\text{inter}} \leq 1$. The program will then calculate the contact friction of the soil and the structural material as shown in the equations below,

$$c_{\text{contact}} = R_{\text{inter}} \times c_{\text{soil}} \tag{7.1}$$

$$(\tan \emptyset)_{\text{contact}} = R_{\text{inter}} \times (\tan \emptyset)_{\text{soil}} \tag{7.2}$$

Unlike geotextile and geomembrane, geogrids have holes, their transversal grids provide a sort of passive resistance to the backfill placed on top of it and the grids also facilitate interlocking between well-compacted backfill soil material to take place (Fig. 7.16). This means the soil particles will move together as if it is one unity and practically allowing no slippage between the soil and the geogrids to take place. The second function of the interface element is to tell the software whether the modeled "geogrids element" is permeable or impermeable. As water can pass through geogrids, the interface element need not be introduced. If the interface element is used, then the interface element must be chosen to be permeable. Based on the above reasons it was decided to take $R_{\text{inter}} = 1$ which is the default value in Plaxis.

- The allowable tensile strength of geogrids, T_{all}, was calculated as follows [8, 12]:

$$T_{\text{all}} = \frac{T_{\text{ult}}}{\text{RF}_{\text{CR}} \times \text{RF}_{\text{ID}} \times \text{RF}_{\text{CBD}} \times \text{RF}_{\text{joint}}} \tag{7.3}$$

where

T_{all} allowable long-term design strength

T_{ult}	short term ultimate strength (breaking strength)
RF_{CR}	reduction factor due to creep
RF_{ID}	reduction factor due to installation damage
RF_{CBD}	reduction factor due to chemical and biological damage
RF_{joint}	reduction factor due to joints or seams

Table 7.2 shows the range of the strength reduction factors for various applications of geosynthetics. The values of RF_{CR} in the table shall be used only when there is no creep reduction or creep isochronous curves available. The low RF_{CR} refers to the applications that have relatively a short service life and/or where creep deformations are not critical to the overall performance of the geosynthetic reinforced structures.

At this project, a design life of 100 years was required. Paralink geogrids with breaking strength of T_{ult} = 300 kN/m from Maccaferri were used. The reduction factor due to creep was taken from the test result as shown in Fig. 7.17 [1]. The chart shows for 100-year design life, the tensile strength will reduce to about 72%, which gives a value of RF_{CR} = 1/0.72 = 1.39. A value of 1.05 was taken for both RF_{ID} and RF_{CBD}. No joint was allowed along with the main grid strength of the geogrids

Table 7.2 Geosynthetic strength reduction factors (modified from [8])

Application	RF_{CR}	RF_{ID}	RF_{CBD}	RF_{joint}
MSE walls	2.0–4.0	1.1–2.0	1.0–1.5	1.0–1.80
Embankments	2.0–3.5	1.1–2.0	1.0–1.5	1.0–1.80
Bearing and foundations	2.0–4.0	1.1–2.0	1.0–1.5	1.0–1.80
Slope stabilization	2.0–3.0	1.1–1.5	1.0–1.5	1.0–1.80

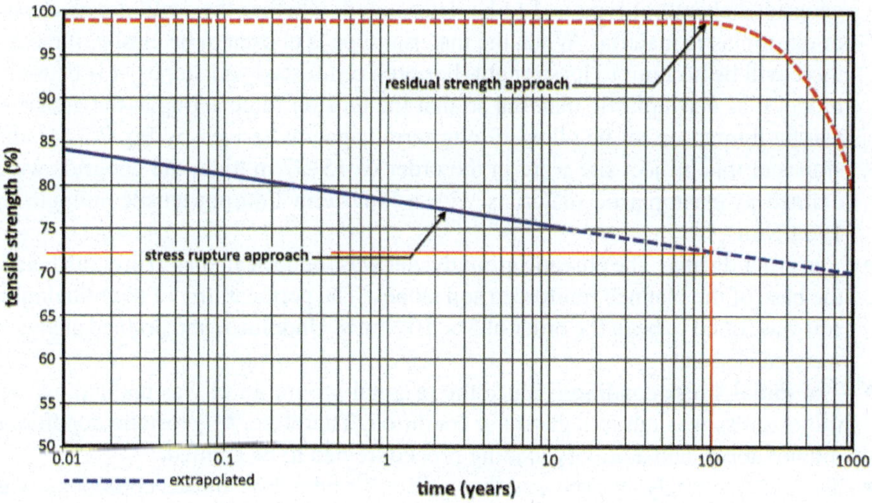

Fig. 7.17 Reduction of strength due to creep of Paralink Geogrids [1]

Table 7.3 Limiting strain of geosynthetics reinforced structures [4]

Application	Limiting strain, ε (%)
MSE walls	3.0–5.0
Embankments	6.0
Bearing and foundations	2.0
Slope stabilization	4.0–5.0

which meant $RF_{joint} = 1.0$. This gave a value of allowable strength, T_{all}, of:

$$T_{all} = \frac{300}{1.39 \cdot 1.05 \cdot 1.05 \cdot 1.0} = 196\,\text{kN/m}$$

- The geogrids' axial stiffness, EA, was then determined as below [4]:

$$E = \frac{\text{Stress}}{\text{Strain}} = \frac{\sigma}{\varepsilon} = \frac{T_{all}}{A} \cdot \frac{1}{\varepsilon} \rightarrow \text{EA} = \frac{T_{all}}{\varepsilon} \tag{7.4}$$

where

E Young's modulus of the geogrid
A cross-section of the geogrid used
ε limiting strain allowed in the geogrids (see Table 7.3)

A limiting strain of 5% was adopted, therefore, EA = 196/5% = 3900 kN/m' (note: it is 3900 kN per m width of the geogrids).

- Geogrids' behavioral type: Apart from determining their tensile stiffness, the geogrids behavioral type had to be determined, whether they will act as elastic or elastoplastic material. When the material type is chosen to be elastic, it means there will be no limit to its strength. For high reinforced soil slopes, it is better to model it as elastoplastic material so that the tension force acting at the geogrids can be limited up to the allowed long-term capacity derived by Eq. 7.3. As the slopes at this project site were in the order of 15–37 m high, the geogrids were defined as elastoplastic materials with a long-term limiting tensile strength of 196 kN/m.
- The groundwater: The phreatic groundwater was found at about 4.0 m depth below the base of the planned reinforced soil slopes. The permeability of the foundation soil was found to be in the order of 5×10^{-5} m/s. Undrained and drained analyses were carried out.
- The initial stress condition: Since the original ground level was undulating, the initial stress was not in a geostatic condition. Therefore, k_o geostatic conditions did not apply, and a gravity loading procedure had to be applied.
- The seismic analysis: The stability analysis under earthquake conditions was analyzed by using a pseudo-static approach with a horizontal seismic coefficient of $k_h = 0.15$ g.

- Further checks on the internal and external stabilities of the reinforced soil slopes were also carried out through limit equilibrium analyses by using Geo5 MSEwall software from Fine CZ.

Several configurations of the slopes were analyzed before the best ones were adopted. Table 7.4. shows the lowest allowable required factor of safety and the overall safety factor for the highest reinforced slopes of 37 m high.

Figure 7.18 shows the predicted overall displacement under static loading. Notice that the maximum predicted settlement at the top surface is around 684 mm while at the foundation level is around 250 mm. This means 434 mm happened at the compacted backfill soil, which in reality it would be self-adjusted during the construction process by the on-site leveling measurement and practices. Therefore, the actual deformation would be only the foundation soil settlement which also partly or mostly would take place during construction. Up to the write-up of this article, no significant settlement was reported. The calculated static safety factor is around 1.73 to 1.80 (see Table 7.4), which is above the 1.50 required by the Indonesian Geotechnical Standard [14].

Figure 7.19 shows the pseudo-static result. The calculated safety factor under seismic conditions is around 1.10–1.12 which meets the minimum 1.10 required by

Table 7.4 The predicted safety factor (SF) of the 37 m high reinforced soil slopes

Loading condition	Minimum SF required	SF obtained by limit equilibrium method	SF obtained by finite element method
Static	1.5	1.73	1.80
Seismic	1.1	1.10	1.12

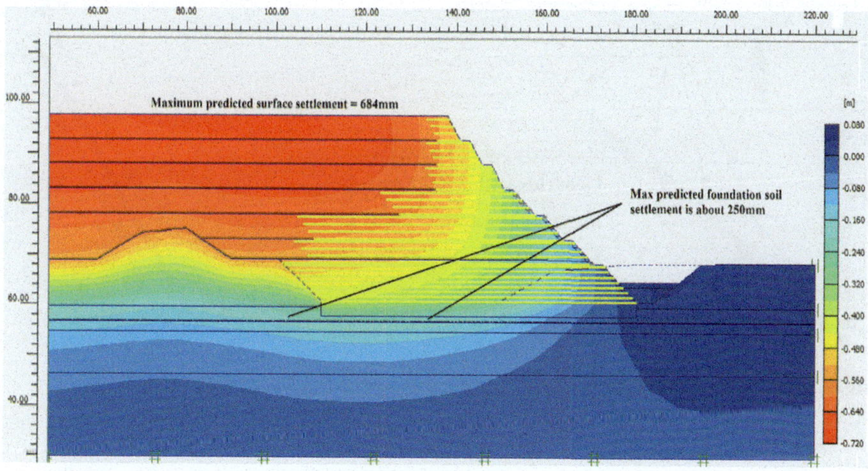

Fig. 7.18 Predicted settlement under static loading

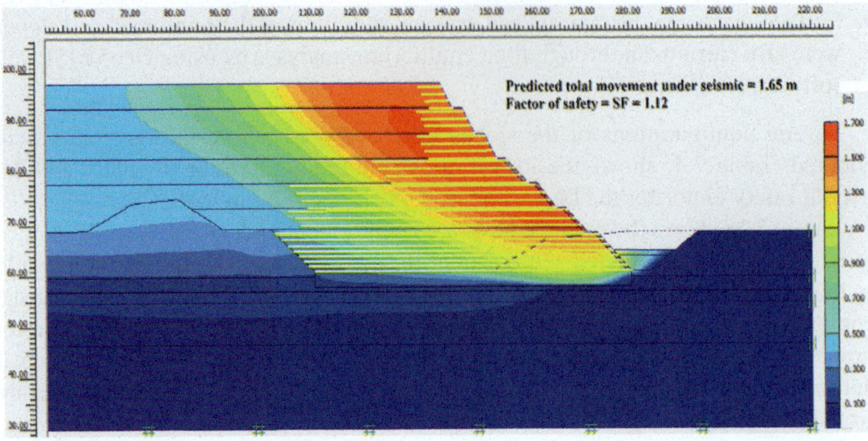

Fig. 7.19 Predicted settlement under seismic loading

the Indonesian Geotechnical Standard [14].. The maximum tensile geogrids force calculated is in the order of 135 kN/m'.

Another consideration during the design process was to prevent the saturation of the constructed reinforced soil slopes. Geosynthetic clay liner (GCL) would cover the top of the reinforced slopes, and adequate surface drainage would be constructed. Sub-soil drainage would also be provided behind the reinforced slopes to drain out seeping rainwater. Figure 7.20 shows the cross-section of the planned soil structures.

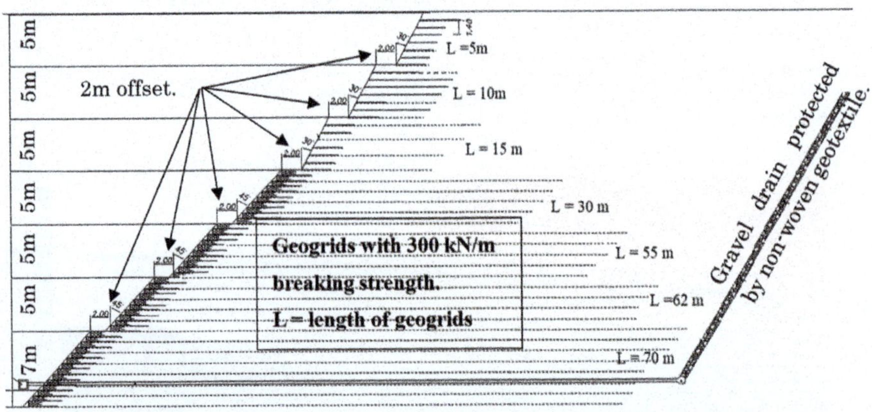

Fig. 7.20 Cross-section of 37 m high planned geosynthetic reinforced soil slope

7.2.4 Construction

The construction of the reinforced soil slopes was carried out in the following procedures:

- Stripping the top 2–7 m of topsoil until hard clay shale was found. Excavation to this foundation level was carried out only in a maximum of 10 m wide strips (Fig. 7.21). The intention was to keep the exposure of the clay shale to the atmosphere as short as possible by quickly covering the clay shale foundation with the first two layers of the reinforced soil structure, hence, keeping the clay shale strength degradation to as minimum as possible.
- Once the hard clay shale foundation layer was found, immediately laid the 1st layer of geogrid, and construct the anchor gabion facing (Fig. 7.22). The geogrid was firmly installed below the gabion base. To prevent intrusion of fined-grained soils into the gabion, the gabion was protected by non-woven geotextile.
- Without any delay, the first layer of backfill material was placed and compacted to the required level of compaction (Fig. 7.23). The compactness of the backfill was monitored by conducting sand cone and dynamic penetrometer tests.
- Geogrid 2nd layer was placed, followed by placing and compacting the 2nd layer of backfill. To cover the clay shale at the base of the reinforced soil wall as fast as possible, the job was carried out day and night (Fig. 7.24). In so doing, the first two layers of the reinforced soil slope system would protect the clay shale foundation against strength degradation.
- At the back of the reinforced soils, a gravel drain (sub-drain) system to protect the compacted backfill against saturation from seeping water was constructed at the same time with the reinforced soil layers (Fig. 7.25). The piping to drain out

Fig. 7.21 Stripping topsoil until hard clay shale foundation (**a**) in a 10 m wide strip (**b**)

Fig. 7.22 Laying geogrid (**a**); preparing empty gabion cage (**b**); stone-filled gabion (**c**)

Fig. 7.23 Immediate placement (**a**) and compaction of first layer backfill (**b**)

the seeping water was covered with a geomembrane. This is to protect against leaking which may cause the saturation of the compacted backfill (Fig. 7.26).

- The above steps were then repeated to the next strips until about 50–100 m in widths were completed. After reaching ± 50 m wide strip, the building up of the

Fig. 7.24 Placement and compaction of second layer geogrid

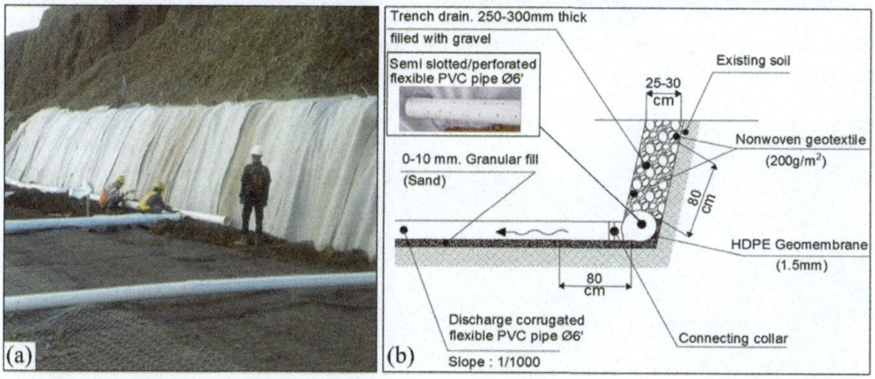

Fig. 7.25 Installation of gravel drains (**a**) and drainage dimension (**b**)

reinforced soil slopes was started layer by layer from the starting point to the end of the strip until reaching the top of the slope (Fig. 7.27).

- The top 10–15 m of the gabion facings were replaced with a wrap-around wire mesh where at its surface a mat is placed to grow grasses (Fig. 7.28).
- Fig. 7.29 shows an overview during the building up of reinforced slopes. Upon reaching the top of the slope, grass seeding was carried out on the facings (Fig. 7.30). The unpaved area was then covered by GCL or low permeability compacted clay layer of at least 50 cm thick, and a surface drainage network was installed to drain out surface runoff during rainy days.

Fig. 7.26 PVC pipe wrapped with geomembranes to prevent leakage (**a**) and finished drainage pipes (**b**)

Fig. 7.27 Laying of subsequent reinforcement layers (**a**), stone-filled gabion facing (**b**), and partially finished reinforced soil slope (**c**)

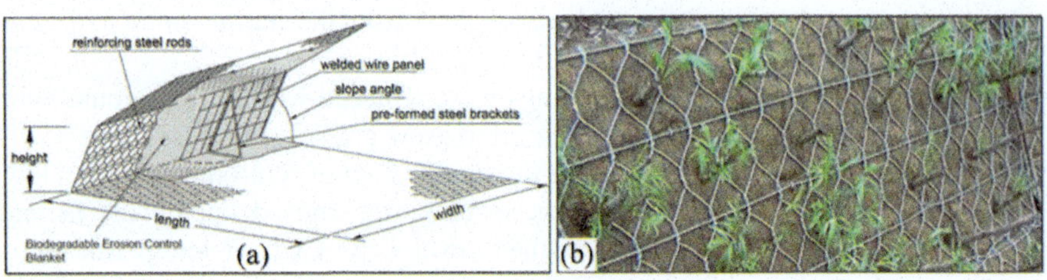

Fig. 7.28 Wrap-around wire mesh facings installed at the top 10–15 m of slopes, schematic detail (**a**), and after installation (**b**)

Fig. 7.29 Construction overview of the finished reinforced soil slopes

7.2.5 *Tests from Flash Flood and Palu Earthquake*

During the process of the construction, before the surface drainage was constructed, a heavy rainstorm hit the area causing a flash flood to flow through one section of a newly completed reinforced soil slope for a few hours. It was fortunate that the reinforced slope stood firm, it suffered only minor defects but the newly grown grasses were wiped out (Fig. 7.31). Repair operations were quickly carried out. By early 2019 some sections, except the highest part, had become green as shown in Fig. 7.32.

On 28 September 2018 when the city of Palu and its vicinity area was struck by a 7.4 Mw devastating earthquake that caused massive flow liquefaction and tsunami, this New Tana Toraja airport, located ± 270 km south of the epicenter, also suffered from the tremor. However, there was no damage reported on the completed high slopes.

7.2.6 *First Plane Landing*

The construction of the airport was completed in 2020. The first plane smoothly landed on August 12, 2020. Figure 7.33 shows the moment of landing plane when it passed the 37 m highest section of the geosynthetic reinforced soil slopes.

Fig. 7.30 One month after seeding

Fig. 7.31 Flash flood over the finished slope (**a**), state of the slope after flash flood (**b**)

7.3 Concluding Remarks

Lessons learned from the presented case study:

Fig. 7.32 State of the slopes in early 2018

Fig. 7.33 First landing on the runway, August 12, 2020

- Clay shale is one of the problematic soil conditions that can go undetected and endanger the stability of geotechnical construction. Considering that the clay shale can lose most of its shear strength in just a few days, construction must be carried out in the fastest time possible. Prolonged exposure of the good clay shale layer to the outside atmosphere must be prevented, and reduced shear strength parameters must be adopted in the design process.
- In the case presented, reduced shear strength parameters of $c' = 20$ kPa and $\phi' = 17^\circ$ were adopted for the first 3 m thick of the clay shale which is far less than its maximum unweathered strength of $c' = 85$ kPa and $\phi' = 41^\circ$.
- Properly designed and carefully executed high slopes reinforced with geogrids and gabion facings can be built safely on clay shales formation.

Finally, a combination of a good understanding of soil behavior, geotechnical knowledge, design experiences, and good construction procedures is required to successfully develop geotechnical structures on problematic soils.

References

1. BBA test certificate: Linear composites' soil reinforcement products, Paralink Geocomposites, The BBA (British Board of Agreement) Approval Inspection Testing Certification, Watford, UK. (2010)
2. Cox, J.B.: The distribution and formation of recent sediments in South East Asia. In: Proceeding 2nd South Asia Conference on Soil Engineering, pp. 30–47 (1970)
3. Gartung, E.: Excavation in hard clays of the Keuper formation. In: Proceeding of Symposium on Geotechnical Engineering, Seattle, Washington (1986)
4. Gouw, T.L.: Importance of elongation factor in determining geosynthetics stiffness for finite element calculation. In: Proceedings of International Conference on Landslides and Slope Stability, Bali, 27–30 Sept. 2015, pp. B8-1 to B8-6 (2015)
5. Gouw, T.L.: Geosynthetics application in Indonesia—case histories. Geotech. Eng. J. SEAGS & AGSSEA **49** (4) Dec 2018 ISSN 0046-5828 (2018a)
6. Gouw, T.L.: Common mistakes in designing MSE wall with finite element method. In: 11th International Conference on Geosynthetics, 16–21 July 2018, Seoul, South Korea (2018b)
7. Irsyam, M.: Geotechnical Engineering Lecture Notes, unpublished, Institut Teknologi Bandung, Indonesia (2012)
8. Koerner, R.M.: Reduction Factors Used in Geosynthetics Design, GSI White Paper 4, p. 13. GII Publications, Folsom, Pennsylvania (2005)
9. Koseki, J., Nakajima, S., Tateyama, M., Watanabe, K. & Shinoda, M.: Seismic performance of geosynthetic reinforced soil retaining walls and their performance-based design in Japan, Theme Lecture. In: Proceedings of the international conference on performance-based design in earthquake geotechnical engineering—from case history to practice, Tsukuba, pp. 149–161 (2009)
10. Koseki, J.: Use of geosynthetics to improve seismic performance of earth structure, Mercer Lecture 2011. Geosynthetics Geomembr. **34**, 51–68 (2012)
11. Kuwano, J., Koseki, J., Miyata, Y.: Performance of reinforced soil walls during the 2011 Tohoku earthquake. Geosynth. Int. **21**(3), 1–18 (2014)
12. Sarsby, R.W.: Geosynthetics in Civil Engineering. Woodhead Publishing Limited, Cambridge, England (2007)
13. SNI 1726:2012: Earthquake Resistance Design Procedures for Building and non-building Structures (in Indonesian), National Standardization Bureau of Indonesia (2012)
14. SNI 8640:2017: Geotechnical Design Requirement (in Indonesian), National Standardization Bureau of Indonesia (2017)
15. Stark, T. D., Duncan, J. M.: Mechanism of strength loss in stiff clays. J. Geotech. Eng. **117** (1) (1991)
16. Tatsuoka, F., Koseki, J., Tateyama, M.: Performance of earth reinforcement structures during the Great Hanshin Earthquake, Special Lecture. In: Ochiai et al. (eds.) Proc. Int. Sym. on Earth Reinforcement, IS Kyushu '96, Balkema, vol. 2, pp. 973–1008 (1997)
17. Wyllie, D.C., Mah, C.W.: Rock Slope Engineering, 4th edn. Spon Press, Taylor & Francis Group, London and New York (2004)

Chapter 8
Slope Stability by Soil Nailing for Road Widening in Hikawa Tunnel, Tokyo, Japan

Juichi Nakazawa

8.1 Outline of Project

The project involved a two-lane road with many curves in a mountainous area of Tokyo. The purpose of this project was to make the road straight for overcoming the problem of heavy traffic. As one of the measures, in the year 1985, it was decided to construct a road tunnel in the hilly part. For connecting the existing road at one exit of the new tunnel, the road has to be widened by cutting the slope (Fig. 8.1). The road passes through an area, on which one side is the river with a vertical cliff of 20 m as seen in Fig. 8.2. On the hilly side, there was a retaining wall of 9.0 m in height made of boulders with gradient of 0.3:1.0. The upper part of the retaining wall had a natural slope with 50 degrees inclination. As shown in Fig. 8.2, at 17 m height, in the flatter part, there was a residential house. In addition, there was an active railway tunnel passing through the hill. As per construction specifications, the challenge of this project was as following.

(1) Service of one of the two lanes must be maintained.
(2) No damage should occur to the existing structures such as the railway tunnel and the residential house.

8.2 Geological Conditions of the Project Site

The distribution of N-value from SPT in the slope is shown in Fig. 8.2, based on the preliminary survey conducted. The geological formations of the site were found to be chart, sandstone, shale of Mesozoic and Jurassic Era. Also, talus deposits were

J. Nakazawa (✉)
Tokyo Engineering Consultants Co. Ltd., Osaka, Japan
e-mail: junakaza@gf7.so-net.ne.jp

H. Hazarika et al. (eds.), *Practices and Trends in Ground Improvement Techniques*, Developments in Geotechnical Engineering,
https://doi.org/10.1007/978-981-19-3322-6_8

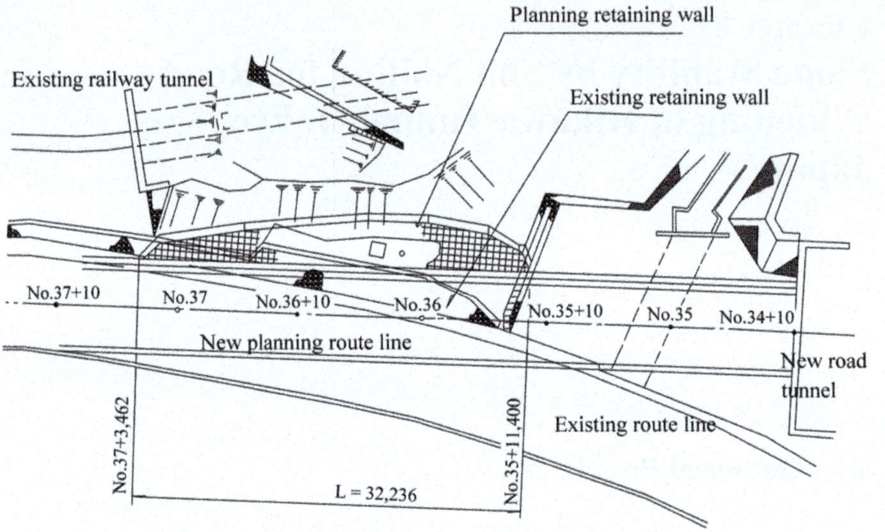

Fig. 8.1 Ground plan [1]

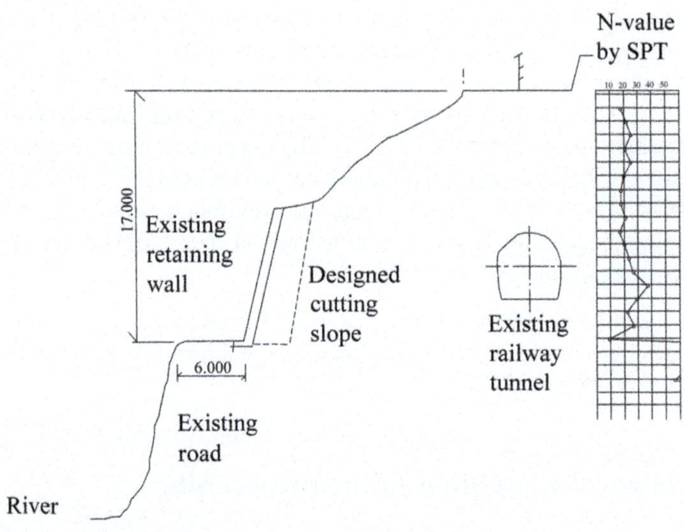

Fig. 8.2 Cross-section and *N*-value [1]

found to be distributed in the upper part. The object slope was in talus deposits that were relatively dense clayey gravel. Geotechnical investigations revealed that the breccia in the deposit was slate, and their particle sizes varied from 2 to 300 mm. In addition, clay lenses and sand lenses were found to be the talus. No groundwater was observed at the site.

8.3 Nailing

Nailing is a popular method for reinforcing weak rock mass. There are varieties of nailing methods. The applied method in this project was based on the Root Piling technique developed by an Italian company [2, 3].

The slope was stabilized by a combination of rebar with 80–100 mm diameter and cast-in-place mortar placed in the drilling hole. Nailing was driven in net-like-pattern inside the ground. The mortar acts as an intervening medium between the nailing and ground and also as reinforcing materials. The mortar should be applied with suitable pressures. The mortar fills the voids between the ground and nailing as well as the cracks within the ground, resulting in the integration of the ground and reinforcing materials. If necessary, expanding agent needs to be added to the mortar. Ground can obtain the tensile strength from the nailing and cast-in-place mortar, which form three-dimensional structure within the ground.

8.4 Procedures

Figure 8.3 shows the arrangement of nailing in the front view. As shown in Fig. 8.3, at each 1.5 m height the slope is cut vertically, and nailing is executed in 10 steps. This procedure is repeated from the top to bottom. Figure 8.4 shows the cross-section of the slope and the details of nails. The execution procedures are as follows:

(1) Slope cutting:

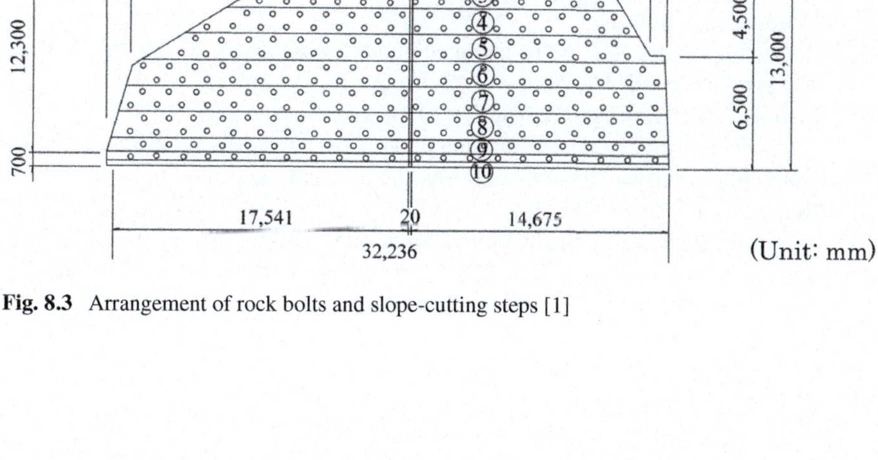

Fig. 8.3 Arrangement of rock bolts and slope-cutting steps [1]

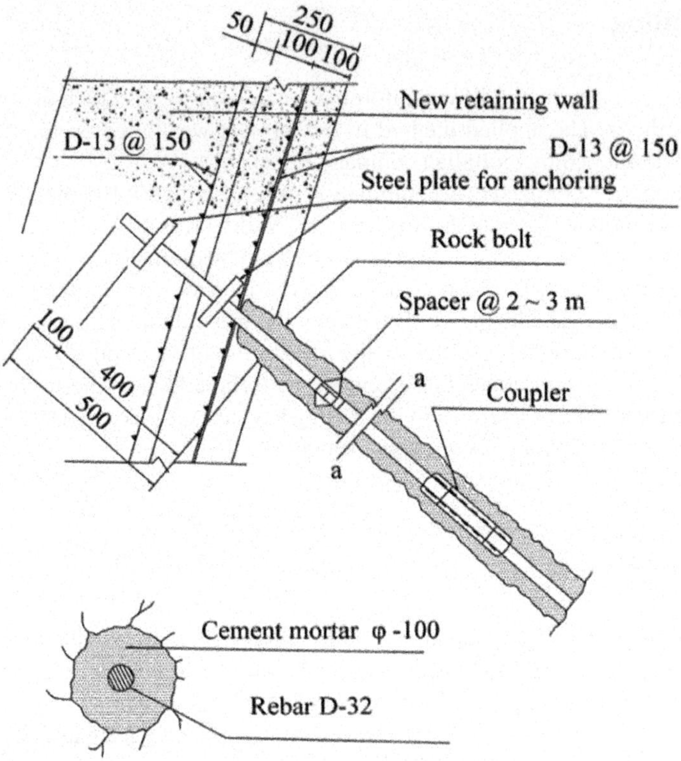

Fig. 8.4 Details of rock-bolt [1]

Slope-cutting work was performed manually because the working space was situated at high level and the area was too narrow for machine work. The vertical height for each step was 1.5 m.

(2) Primary shotcrete:

Primary shotcrete for the slope lining of thickness 10 cm was done immediately after cutting.

(3) Drilling for nailing:

Drilling work was done using compressed air. Drilling bit of diameter 125 mm was used. The diameter of casing was 118 mm.

(4) Installation of nailing and injection of mortar:

After drilling work is completed, D-32 type nail was installed. Then, concrete mortar was injected between the nail and the drilled hole.

(5) Removal of casing tube:

The casing tube should be pulled out in stages during the injection of mortar until the entire drilling hole is filled with mortar.

(6) Secondary shotcrete and placement of rebar:

A secondary shotcrete of thickness of 10 cm was placed on the top of the primary shotcrete surface. D-13 type rebar was then placed at 150 mm square spacing on the secondary shotcrete surface and the rebar was fixed using fixing plate as shown in Fig. 8.4.

(7) Construction of leaning-type retaining wall:

After slope-cutting works are completed by repeating the above-mentioned procedures, the leaning-type retaining wall was constructed on the reinforced slope in stages, starting from the bottom of the slope.

8.5 Stability Analysis of the Slope and Monitoring

Prior to proceeding with this project, deformations and stress distributions due to slope cutting were analyzed by performing slope stability analysis using FEM. Table 8.1 shows the values of different parameters of the ground. Failure condition was judged by Mohr–Coulomb's failure criteria. The analysis showed that until 6th step, there was no failure. However, from the seventh step until the tenth step, two failure elements at the middle part of the slope were recognized. The overall stability of the cutting slope was considered to be safe because no. of failure elements were too few. Therefore, it was decided that the slope should be monitored by measuring and observing the two unsafe locations. The axial forces generated in the rebar were estimated to be 1.0 tons to 5.0 tons per rebar. Since the allowable tensile strength of rebar was calculated to be 14.3 tons per rebar, the safety factor of rebar was about 2.6.

As stated before, near the construction site, there were an active tunnel and a private house. The tunnel is situated on the hillside of the construction site, at a horizontal distance of about 15 m away from the cutting slope. Observation and monitoring management of the cutting slope was undertaken for aiding safety and evaluation of the cutting slope. Measuring items were as mentioned below:

(a) Two borehole inclinometers to monitor underground displacement.
(b) Placement of surveying point to monitor ground surface displacement.
(c) Measurement of axial load of nails.

Table 8.1 Soil parameters for simulation

Items	Parameters
Unit weight	2.00 t/m^3
Coefficient of earth pressure at rest	0.50
Modulus of elasticity	10,000 kN/m^2
Poisson's ratio	0.333
Cohesion	20 kN/m^2
Angle of internal friction	35°

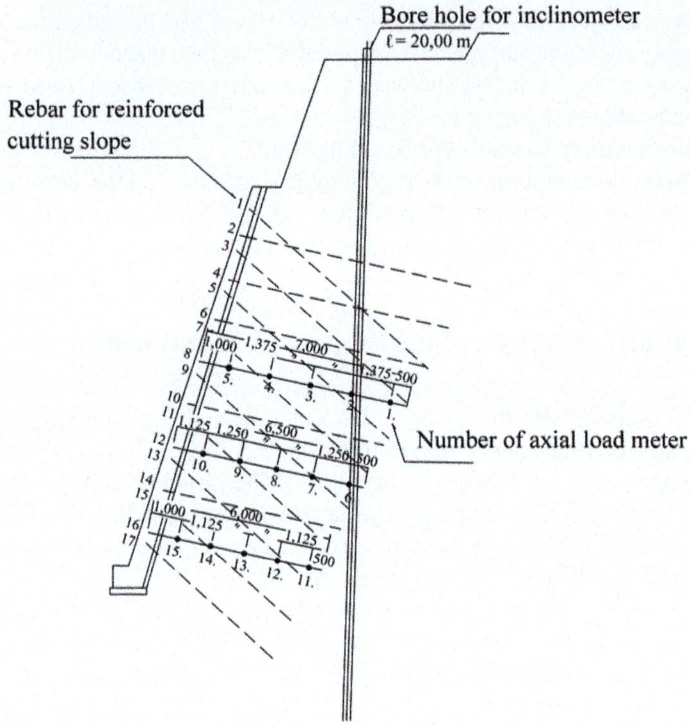

Fig. 8.5 Arrangement of measuring instruments [1]

The positions of the borehole inclinometers and axial load gauges are shown in Fig. 8.5.

8.5.1 Displacement of Cutting Slope

Representative results from the measurements are shown in Figs. 8.6 and 8.7. Figure 8.6 shows the displacements inside the slope obtained from the borehole inclinometers. From these results, the horizontal displacements toward the valley are observed with the progress of the slope cutting. The final horizontal displacement at the top of slope was measured to be 16–17 mm. Based on the slope height (about 17 m), the displacement rate was calculated to be 1/1000. At the end of the slope-cutting work, changes in slope displacement were not observed. With the construction of the leaning-type retaining wall, gradually the rate of change of displacement decreased and the slope returned to the stable condition.

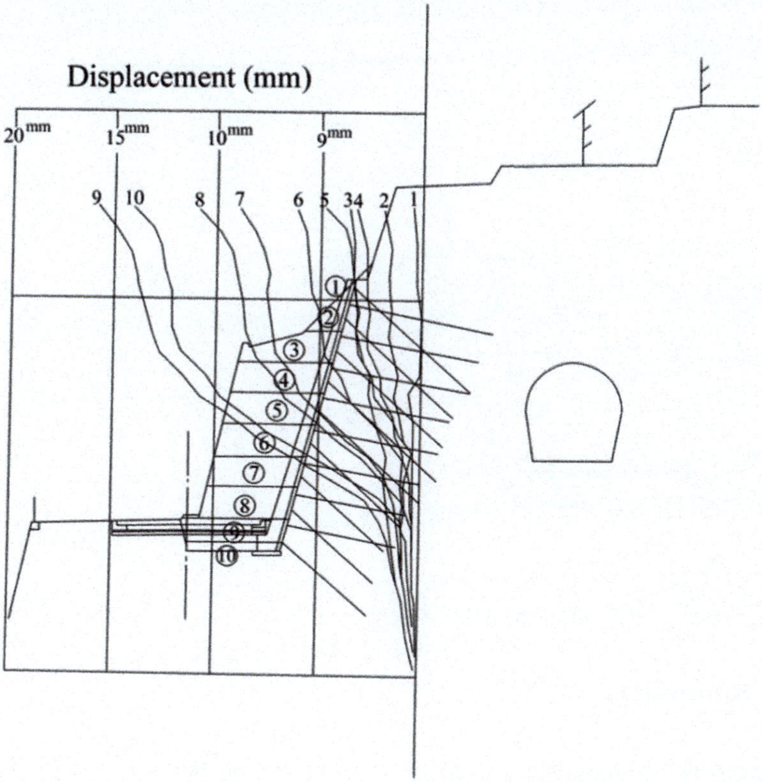

Fig. 8.6 Displacements by slope-cutting step [1]

8.5.2 Axial Load

The maximum values, which were measured by the axial load gauges, were 5 tf at location 8, 6.5 tf at location 12, and 2 tf at location 16. Since the allowable tensile strength of rebar was 14.3 tf, these values were considered to be safe. The distribution of axial load for a rebar (Fig. 8.7) was larger at the center and smaller at both the end. These results show that the nailing acted to reinforce the cutting slope effectively. As a whole, the axial loads of rebar converged to constant values at the end of cutting work, and the slope stabilized with the construction of the leaning-type retaining wall.

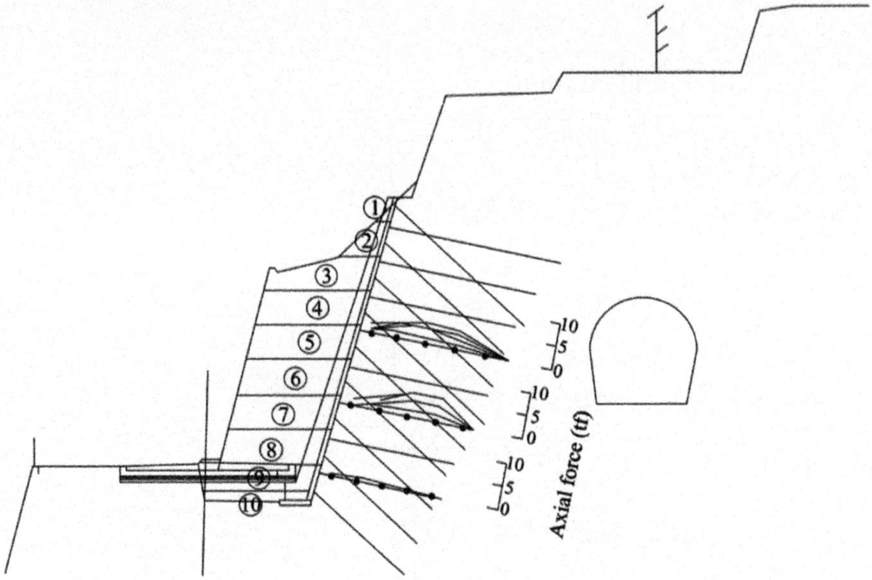

Fig. 8.7 Axial load by slope-cutting step [1]

8.6 Summary

While cutting and reinforcing the slope, and then moving to the next step of cutting, this nailing technique was evaluated to be very successful. During this project, no damage was observed to the existing railway tunnel and the existing house. The state of the roads in the year 2016 was shown in Fig. 8.8.

Fig. 8.8 Present state after construction

References

1. Fujii, I., Igarashi, Y., Fukuoka, H., Nishihama, J., Nakazawa, J.: Slope protection work with nailing earth reinforcement. Technical Reports of Konoike Construction Co. Ltd., vol. 59, pp. 22–30 (In Japanese) (1986)
2. Lizzi, F.: Reticulated Root Piles to correct landslides, ASCE Convention, Chicago, Preprint 3370 (1978)
3. Lizzi, F.: The pali radice (root piles), Symposium on Soil and Rock Improvement Techniques Including Geotextiles, Reinforced Earth and Modern Piling Methods, Bangkok, Paper D-3 (1982)

Chapter 9
Use of Coir Geotextiles in Erosion Control, South-West India

Krishnan Balan, P. K. Jayasree, and C. R. Devaraj

9.1 Introduction

Geotextiles with natural fibers like coir, jute, sisal, hemp, etc. are used in civil engineering applications. Mainly they are used for erosion control applications. The main advantage of geotextiles with natural fibers is that they are ecofriendly and biodegradable. They can be successfully used in limited life applications of geotechnical engineering. The life of coir geotextiles is longer, more than five years, as compared to other natural geotextiles. The inherent tensile strength, drainage and filtration properties of coir geotextiles can be made use of in various civil engineering applications.

9.2 Erosion Control Appications

Natural fiber geotextiles such as coir and jute are widely used for erosion control. They are ideal for controlling soil erosion due to rainfall and runoff along slopes. The biodegradability nature of natural fiber geotextiles is successfully utilized in this application. Soil erosion along slopes is ultimately controlled by the vegetation on it

K. Balan (✉)
Rajadhani Institute of Engineering and Technology, Attingal, Trivandrum, Kerala, India
e-mail: drkbalan@gmail.com

P. K. Jayasree
College of Engineering Trivandrum, Trivandrum, Kerala, India
e-mail: jayasreepk@cet.ac.in

C. R. Devaraj
Charankattu Coir, Shertallay, Kerala, India
e-mail: devaraj@naturalworld.in

H. Hazarika et al. (eds.), *Practices and Trends in Ground Improvement Techniques*,
Developments in Geotechnical Engineering,
https://doi.org/10.1007/978-981-19-3322-6_9

and natural fiber geotextiles provide a micro-climate for the speedy germination of vegetation on virgin/natural slopes. Coir geotextiles prevent the surficial soil erosion till the vegetation establishes.

Bureau of Indian Standards has brought out Indian Standard Code of Practice for erosion control applications of coir and jute in highway and railway embankments (IS: 15872, 2009). Indian Railway also accepted coir and jute as erosion control material along slopes (Guideline No. GE: G-4).

Of the natural fiber geotextiles available on market, coir geotextiles have greater longevity. Both woven mesh mattings and non-woven blankets are used for erosion control. Though both types are ideal for almost all occasions, non-woven blankets are ideal where there is a great potential for rainfall erosion than runoff erosion. The properties of coir geotextiles used in erosion control are summarized in Tables 9.1

Table 9.1 Standard Properties of woven coir geotextiles

S. No	Characteristic	Grade			Method of test
1	Mass/unit area, g/m^2, Min	400	700	900	IS: 15868
2	Thickness at 20 kPa, mm, Min	6.5	6.5	6.5	IS: 15868
3	Break load, dry (kN/m), Min (a) Machine direction (b) Cross machine direction	7.00 4.00	8.5 8.0	15.0 8.0	IS: 13162 (Part 5)
4	Break load, wet (kN/m), Min (a) Machine direction (b) Cross machine direction	3.0 2.0	7.0 4.5	12.5 5.0	IS: 13162
5	Peak load, dry (kN/m), Min (a) Machine direction (b) Cross machine direction	7.5 4.0	9.0 8.0	18.0 9.0	IS: 13162
6	Peak load, wet (kN/m), Min (a) Machine direction (b) Cross machine direction	3.0 2.0	8.5 5.5	15.0 6.0	IS: 13162
7	Trapezoidal tearing strength (kN) at 25 mm gauge length, Min (a) Machine direction (b) Cross machine direction	0.18 0.15	0.35 0.30	0.50 0.35	
8	Mesh size, mm, Max	20.0 × 16.75	7.50 × 7.30	4.2 × 5.1	IS:15868

Table 9.2 Properties of stitched coir erosion control blanket

Description	Values
Fiber	100% coir
Width and length	2.4 m × 42 m
Unit weight	450 gsm
Top and bottom PP netting	14 × 14 mm; 4.8 gsm
Stitching thread	HDPE monofilament, UV stabilized
Tensile Strength ISO—10319	104 N/m
Maximum permissible shear stress	1.10 gsm

and 9.2.

The success of natural fiber geotextiles in erosion control lies in the correct laying of the material in the field. Some case histories on use of coir geotextiles in erosion control are presented below.

9.3 Case Studies on Mine Waste Dumps

Erosion control of mine waste dumps using both woven or stitched non-woven coir blankets was executed in various mines in India, such as Sessa Goa Iron ore mine in Goa, Roida Iron ore mine in Orissa, and Rampura Agucha zinc mines near Jaipur in Rajasthan. Both rainfall and runoff have equal influence on the stability of the mine waste dumps, by creating deep gullies along the surface.

The usual failure pattern of mine waste dump is shown in Fig. 9.1. The wooden or steel spikes used to anchor the geotextile to ground were of varying lengths. Spikes

Fig. 9.1 Failure pattern of mine waste dumps

Fig. 9.2 Creation of bund on top of the slope and trench to anchor the coir geotextiles

of about 1.00–1.20 m were used in the top region and reduced lengths of up to 0.30 m were used in the bottom region. Top anchoring of the geotextiles was made beyond the width of the gully on surface of the slope, the width of the gully usually varied from 1.0 to 1.5 m.

Geotextiles were anchored at a distance of more than 2.0 m from the top edge of the slope. A small bund is constructed on the top of the waste dump near the top edge of slope, so that no direct flow of water from the surface of the dump into the slope can be prevented (Fig. 9.2). Woven coir geotextile was used in Sessa Goa Iron ore mine waste dumps. Woven coir geotextile in combination with coir pith was used in the zinc mine in Rajasthan. Stitched coir erosion control blanket was used in Iron ore mine in Orissa.

Both woven meshes (maximum opening size limited to 2.5 cm × 1.25 cm) and non-woven blankets having minimum of 0.1 kN/m strength can be used to control the surficial erosion along mine waste dump slopes. The laying and anchoring of the geotextile, as a preventive measure against the surface failure of slope, is the most important criteria, for the better performance of the natural fiber geotextiles, as far as mine waste dump protection is considered. Laying of 0.5 cm × 1 cm mesh opening size coir geotextile at the Rampura Agucha zinc mines near Jaipur in Rajasthan is shown in Fig. 9.3.

Bamboo spikes were used to anchor the coir mesh mattings to the ground. Steel spikes were used only at the joints and that too minimum possible, so that the system will be of natural material alone. Since the top soil in the waste dump consists of soil from far below surface of the ground, it will not have any nutritional factors for the growth of vegetation. Hence to increase the water retention of the soil and also to create a macro climate for the growth of vegetation, coir pith at the rate of 50 to 100 gm per square meter of the geotextile treated area was spread over. Vetiver grass was used as vegetation along the treated slope. Controlled watering was done over the area to have proper growth of the vetiver grass.

Fig. 9.3 Laying of coir mesh mattings over zinc mine waste in Rajasthan

Stitched non-woven coir blankets of 450 gsm were used in Roida Iron ore mine waste dumps in Orissa (Fig. 9.4).

Of the woven and stitched coir blanket, stitched coir blanket performed better than the woven in protecting mine waste dumps in arid climatic conditions prevailing in Orissa and Rajasthan.

Fig. 9.4 Laying of stitched coir blanket at Roida Iron ore mine dumps

9.4 Slope Protection with Coir Geotextiles and Hydro-seeding

A new hill city project based on the principles of New Urbanism, near Pune was taken up by Lavasa, the real estate division of Hindustan Construction Company, Bombay. An ecofriendly town planning is proposed in this project. The city is spread over 100 km^2 almost one-fourth size of Mumbai. The city was set amidst seven hills of Sahyadri Mountains with 60 km lake front along Warasgaon Lake. The principle for sustainable growth at Lavasa adopted a two-pronged strategy; viz, protecting the existing natural habitat as it is and further enhancing the habitat through hydro-seeding, mass plantations, and beautification of ravines. No effort has been spared to ensure that Lavasa maintains nature's ability to replenish and renew its resources.

Hydro-seeding over CCM woven and stitched non-woven coir geotextiles was conducted and a comparative performance evaluation was done in soft lateritic soil slopes. Figure 9.5 shows the hydro-seeding over the control section. Figure 9.6 shows hydro-seeding over woven coir geotextiles. It has been shown that hydro-seeding over coir geotextiles is more effective for speedy vegetation growth compared to that seeding without geotextiles. Among woven and non-woven, the performance of non-woven coir geotextiles is found to be more effective than woven. More displacement and concentrated patching of hydro-seed were found in woven geotextiles, whereas a uniform distribution of hydro-seed could be observed in non-woven coir geotextiles after 2 weeks of application of hydro-seeding (Fig. 9.7). A comparative evaluation of vegetation growth on hydro-seeding alone and CCM geotextiles with hydro-seeding is shown in Fig. 9.8 [1, 2].

Fig. 9.5 Hydro-seeding in control section

Fig. 9.6 Hydro-seeding over woven coir geotextiles

Fig. 9.7 Uniform vegetation and mulching over stitched coir blanket and wash out and concentration of mulch in woven coir geotextiles

(a) Area without geotextiles (b) Area with coir geotextiles

Fig. 9.8 Vegetation growth in hydro-seeded area

9.5 Slope Stabilization of High-Altitude Cricket Stadium

The first high-altitude cricket stadium in India was constructed in Mahendragri by the District Cricket Association of Wayanad District in Kerala, India. The southern side of the proposed stadium has a retaining wall to retain the filled-up soil. The height of the retaining wall was 9 m. Above the retaining wall; a further filling of 5 m is needed. The problem was referred to the authors at this stage. Further filling was made with the locally available soil of silty sand with clay (forest loam). The slope of the filled embankment was 1:1.2. In the natural condition, the soil at the site is very intact, but the excavated soil will become slushier as the water content increases. Hence proper drainage of the surface of the ground is required.

The finished surface of the stadium will have a marginal slope to all the sides from the center potion so that rain water will not be stagnated in the stadium. Concrete edge drain to collect the rain water from the surface of the play area is constructed all around the stadium. The slope of the drain has been made in such a way that the storm water is disposed off safely, on either side of high retaining wall, without coming on the top of the sloping face of the filled-up soil. Edge drains are to be constructed at about 2 m away from the top edge of the filled-up slope. Protective nettings will be provided at about 1.50 m from the top edge of the slope, i.e., 0.50 m away from the outer edge of the drain.

Owing to the above procedure, the filled-up slope will experience only the direct downpour of the rain on the slope surface during rainy season. Being a cricket stadium, at high altitudes, paving of the slope with rock boulders is not recommended considering the aesthetic and to maintain the natural appearance, hence a soil–bioengineering solution was adopted. In order to protect the loosening of the surface soil along the slope during direct impact of rain water and also to prevent the subsequent downward movement of soil particles, it has been decided to use Erosion Control Blankets (ECB's) made of coir fibers. Coir Erosion Control Blankets will protect the soil along the slope surface against splash and surface erosion.

Since the slope is to be protected against the rain water splash erosion, conventional woven or mesh coir mattings are not effective and hence it has been decided to use CCM Coir Blankets (Non-woven) of 450 gsm. The same blanket has been used as a filter media along the trench at the base of the slope, i.e., at the top of the retaining wall.

Coir ECB used at the site consists of non-woven coir stitched with HDPE monofilaments, UV stabilized, and supported on the top and bottom with thin PP netting. The properties of the coir ECB used are given in Table 8.2.

A schematic diagram of the retaining wall and the filling is shown in Fig. 9.9. The filling has a length of about 900 m. In order to divert the surface water coming from the ground area running over the slope, a concrete drain was constructed on the outer periphery of the playground area and about 2.0 m away from the top edge of the slope. The filling was made with proper compaction control and each lift is limited to 30 cm. It was assured that 95% of the proctor density is achieved through compaction. After reaching the required top level of the ground, uncompacted soil

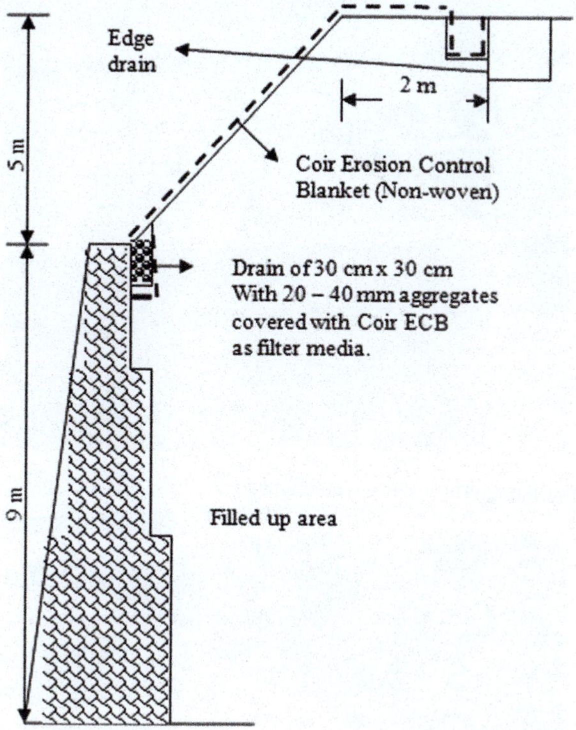

Fig. 9.9 Schematic diagram of the proposed slope protection measures

along the slope was trimmed off (Fig. 9.10). Geotextiles were anchored at the top in trenches taken very near to the concrete drains (Fig. 9.11). A trench, similar to the one at the top is provided at the base of the slope, i.e., just at the top level of the retaining wall. Coir geotextiles are extended to these trenches, properly anchored, and usually back filled with soil. However, in order to avoid the overflow of surface water coming from the slope surface over the retaining wall, the bottom trench has been filled with aggregate of sizes ranging from 20 to 40 mm, to act as drain covered with coir geotextiles. Good quality stones from the cutting area graded to the above size are used for this purpose. The drain has been provided with sufficient slope for easy drainage of water.

Slope treated with coir erosion control blanket has been further treated with planting of Bermuda grass. Since the slope is visible from the entry road itself, it was decided to have ornamental grass rather than thickly growing bush-type (Vetiver) plants. The grass planted over the geotextile helped in preventing the surface erosion.

Stitched coir erosion control blankets were laid on the prepared slope as shown in Fig. 9.12. The grass got good growth within a period of one week and total greenery was visible on the slope from the entry side itself. Vegetation on the slope after 5 months of installation is shown in Fig. 9.13. This work has been executed in 2013.

Fig. 9.10 Prepared slope surface above retaining wall

Fig. 9.11 Laying and fixing of the coir erosion control blanket

Fig. 9.12 Slope above the retaining wall is treated with coir erosion control blanket

Fig. 9.13 Vegetation is in full along the slope after 5 months

9.6 Conclusions

Coir geotextiles are available in the form of woven, stitched non-woven blankets for various civil engineering end uses. Erosion control using coir geotextiles is well established with so many field trials all over the world and both woven and non-woven coir geotextiles are ideal for surficial erosion control applications. Among the available natural geotextiles in the market, coir has got the maximum life and hence can withstand more than three seasons in any extreme climatic conditions for erosion control applications.

All the reported cases are successful and coir geotextiles were intact in the structure till the intended design purpose is served thus showing that the biodegradation part of natural fibers can be best utilized in structures.

References

1. Balan, K.: Application of Coir Geotextiles in Slopes at Lavasa Pune. Proc. National Conference on Geosynthetics, CBIP, New Delhi (2010)
2. Balan, K.: An overview of case and model studies with coir geotextiles. Proc. 5th Asian Regional Conference on Geosynthetics, Bangkok, Thailand (2012)

Chapter 10
Applications of Waste Tires for Protection of Embankments in Northern and Northeastern Japan

Hemanta Hazarika, Kentaro Kuribayashi, Shuichi Kuroda, Kazuya Yasuhara, and Ashoke K. Karmokar

10.1 Introduction

In recent years, engineers and researchers around the world have been developing different techniques to utilize the waste tire (scarp tire) derived materials as geomaterials to improve ground. In Japan, research on such geomaterials started in the early 2000s. To date, sustained efforts by Japanese researchers have led to many pioneering and innovative recycling techniques of scrap tires in civil engineering applications [2, 3, 4, 6, 8, 9, 16, 17]. A comprehensive collection of papers on worldwide research and applications of waste tires can be found in [7].

Applications of the Scrap tire-Derived GeoMaterials (SDGM) can be broadly classified into two categories: (1) Stand-alone application (use of tires and tire-derived materials such as tire shreds, tire chips, in their original form) and (2) Composite application (tire chips mixed with soils and/or other materials). Both categories of applications have many potential benefits. Advantages of using such SDGM in

H. Hazarika (✉)
Kyushu University, Fukuoka, Japan
e-mail: hazarika@civil.kyushu-u.ac.jp

K. Kuribayashi · S. Kuroda
Eight-Japan Engineering Consultants Inc., Okayama, Japan
e-mail: kuribayashi-ke@ej-hds.co.jp

S. Kuroda
e-mail: kuroda-shu@ej-hds.co.jp

K. Yasuhara
Ibaraki University, Hitachi, Japan
e-mail: kazuya.yasuhara.0927@vc.ibaraki.ac.jp

A. K. Karmokar
Bridgestone Corporation, Tokyo, Japan
e-mail: ashoke.karmokar@bridgestone.com

H. Hazarika et al. (eds.), *Practices and Trends in Ground Improvement Techniques*, Developments in Geotechnical Engineering,
https://doi.org/10.1007/978-981-19-3322-6_10

civil and geotechnical engineering practices are manifold: lightweight, vibration-absorbing capacity, compressibility, permeability, and thermal conductivity.

This chapter describes two case studies on the usage of scrap tires. One is the forensic study of an existing retaining wall made of waste tires, which did not suffer any damage during the 2011 Off the Pacific Coast of Tohoku Earthquake and subsequent tsunami. Performances of the tire retaining wall were first evaluated by conducting forensic investigations, followed by laboratory investigations and numerical simulation. The other case study is related to the applications of shredded STDM (called tire shreds) for drainage enhancement of a highway embankment. A full-scale field trial in Hokkaido, Japan is described here, in which tire shreds were used as drainage layer in an embankment under jurisdiction of East Japan Nippon Expressway Co., Ltd., Japan.

10.2 Applications of Whole Tires in Retaining Wall Construction

A retaining wall (made of recycled tires) miraculously survived the earthquake and tsunami disasters during the 2011 off the Pacific Coast of Tohoku Earthquake in Okirai area, Iwate prefecture, Japan (Fig. 10.1a). Ironically, this tire retaining wall is located just about 150 m away from a completely collapsed sea wall (Fig. 10.1b). Due to tsunami runoff, a factory building situated on the backfill ground of the retaining wall was severely damaged, and a natural slope nearby the tire retaining wall was heavily eroded. To make clear, why the strong earthquake of magnitude 8.9 could not inflict any damage to this tire retaining wall, detailed analyses were performed based on the results of forensic investigations, laboratory tests, and numerical simulations.

10.2.1 Forensic Investigation and Results

The whole of Okirai area of Iwate prefecture, where the tire retaining wall is located, was completely inundated by the tsunami during the 2011 Great East Japan Disaster. The causalities in this area include 66 deaths and 30 missing people. As seen in Fig. 10.1b, the concrete sea wall along the coastal line completely collapsed. As a result, the tsunami run-up washed away most of the wooden buildings in the area. The tsunami run-up was up to the third floor of the Okirai elementary school, which is located 200 m from the coastline. The tsunami run-up height in this location was found to be 16 m by the RTK- GPS survey conducted by the Hazarika et al. [5]. As a part of the investigations on the retaining wall, GPS surveying, portable dynamic cone penetration (PDCP) test, surface wave exploration, and micro tremor measurement were conducted on the backfill as well as the toe of retaining wall.

Figure 10.2 shows the plane view of the tire retaining wall. The figure also shows the locations of the various field investigations that were conducted. Disturbed

(a) Undamaged tire retaining wall

(b) Completely damaged concrete sea wall

Fig. 10.1 State of a tire retaining wall and a concrete sea wall after the 2011 Great East Japan Disaster (From [5])

soil samples were also collected from the three locations (No.1, No.2, and soils within tires) and laboratory investigations were carried out to determine the soil characteristics.

PDCP test is recognized as one of the standard methods for obtaining dynamic characteristics of soils at sites by the Japanese Geotechnical Society (JGS 1433). In this test, a drop hammer weighing 5 kg was allowed to fall through a rod from 50 cm height, which enables the cone attached at the toe of the rod to penetrate into the ground. The number of blows (Nd) to penetrate every 10 cm of the ground was measured. Nd values were then converted to the *N*-values of the standard penetration test, using the relationship proposed by Okada et al. [13] for sandy soils. The location

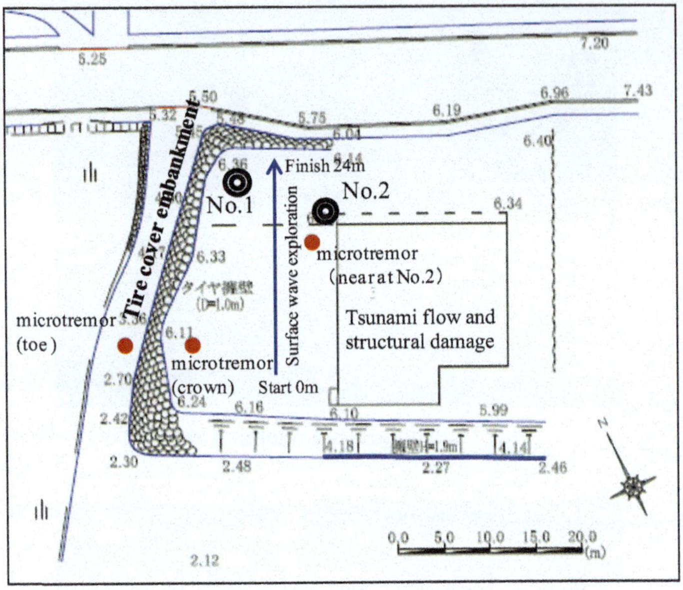

Fig. 10.2 Plan view of the wall and survey locations

of the groundwater table was judged from the water mark on the rod immediately after withdrawal of the rod.

Surface wave exploration is a convenient method to obtain the shear wave velocity distribution (within the ground up to a depth of 10 m), which can measure and analyze the transmission of the surface wave (Rayleigh wave) that transmits near the ground surface. In this method, a wave is generated by striking the ground surface with a hammer. The generated wave propagates according to the surface and subsurface ground conditions. During the investigations, to obtain the characteristics of the ground layer indirectly, surface wave exploration was carried out together with the PDCP test.

On the other hand, micro tremor measurement is another powerful tool to estimate the ground motion characteristics, amplification of ground motion in the soil deposits, microzonation, and dynamic behavior of existing service structures. Micro tremor observations described here were carried out by portable micro tremor equipment (type New PIC). Measurements (two horizontal and one vertical component) were conducted in the velocity mode, which was recorded by the sensors. H/V Ratio was then calculated based on the smoothened ratio of horizontal to vertical Fourier spectra of the micro tremor data. The amplitude ratio calculated in this study was based on the method proposed by Nakamura [15]. The value corresponding to the peak represents the predominant frequency of the ground motion.

Distribution of shear wave velocity within soil was obtained from the results of surface wave exploration method conducted on the backfill soil, which is shown in Fig. 10.3. As seen in the figure, beyond the depth of 10 m, S-wave velocity is

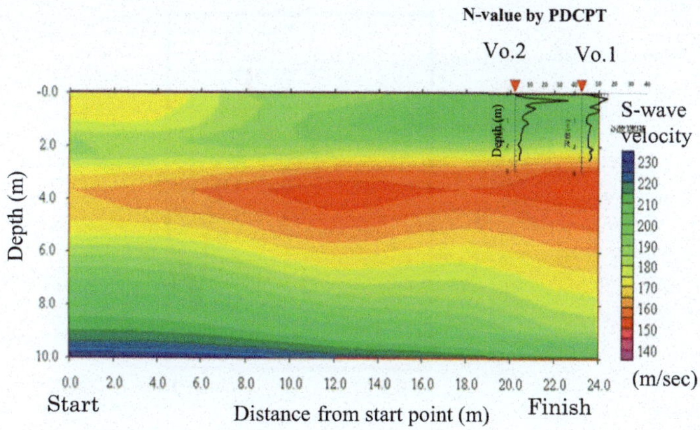

Fig. 10.3 Distribution of shear wave velocity and PDCPT values (From [5])

greater than 220 m/s implying a hard stratum near the sloping side. Also, stratum with 150–200 m/s of S-wave velocity exists within the depth of less than 10 m. Since the average height of the retaining wall is 3.2 m with the maximum height of about 4 m, it can be said that the backfill soil was in the loose state. Figure 10.3 also shows converted N-value obtained from the PDCP test. Converted N-value and S-wave velocity in general are showing the similar trend. The converted N-values within the depth of 0–70 cm are high, implying that close to the backfill surface, the soil density was high. The higher density near the surface indicates the influence of cyclic load imparted to the backfill soils by the parked vehicles, since the yard was used as a parking lot. On the other hand, converted N-value within the depth of 1–1.5 m is about 5, implying that the backfill soil was in a loose state.

Figure 10.4 shows the results of the micro tremor measurements conducted at the top as well as at the bottom of the wall. At location No. 2, the predominant frequency of H/V ratio of the soil deposits is 3.1 Hz (Fig. 10.4a). Spectral ratio between the top and the bottom of the sloping side was found to be 2.1 Hz (Fig. 10.4b). The shear wave velocity from Fig. 10.3 was found to be approximately 200 m/sec beyond 10 m depth. The predominant frequency of the site was found to be 3.8 Hz. Therefore, it can be said that the results of the microtremor measurements and the surface wave exploration are in good agreement.

10.2.2 Laboratory Investigations Using Element Testing and Model Testing

Friction between the tires is considered to be important factor, which resists the slip failures of embankments during an earthquake. In addition, the ingrained viscoelasticity peculiar to tires can be expected to have played a role in damping the seismic

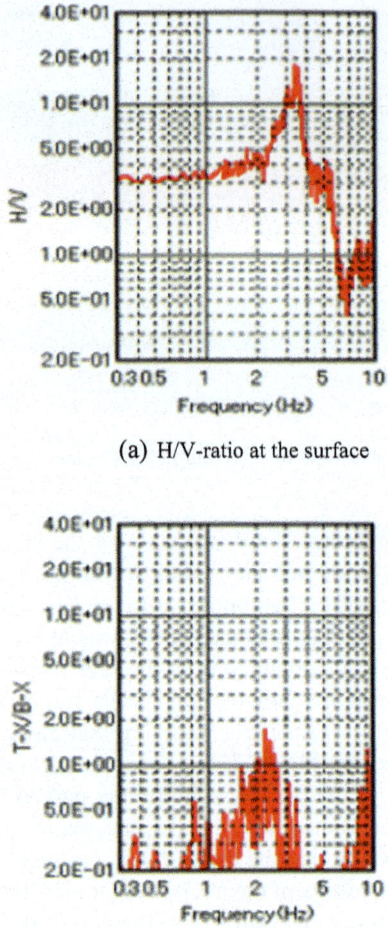

(a) H/V-ratio at the surface

(b) Transfer function

Fig. 10.4 Results from microtremor observations (From [5])

energy. However, there is no previous research clarifying such characteristics of tires. Therefore, a series of model experiments were conducted on specimens imitating the tire retaining wall at the site. The frictional and damping characteristics of the tires were evaluated based on the results of the experiments.

Static and cyclic direct shear tests were performed using a lower movable direct shear testing apparatus containing large size shear box as shown in Fig. 10.5. The target specimen was prepared using eight model tires with a diameter of 85 mm and placed above and below the shear box as shown in Fig. 10.5. The hollow part of the model tire was filled with soil, in the same way as the prototype tires at the site. The same soil, which was used in filling the prototype tires at the site, was used as filling material. Figure 10.6 shows the grain size distribution curve of the filled soil sample

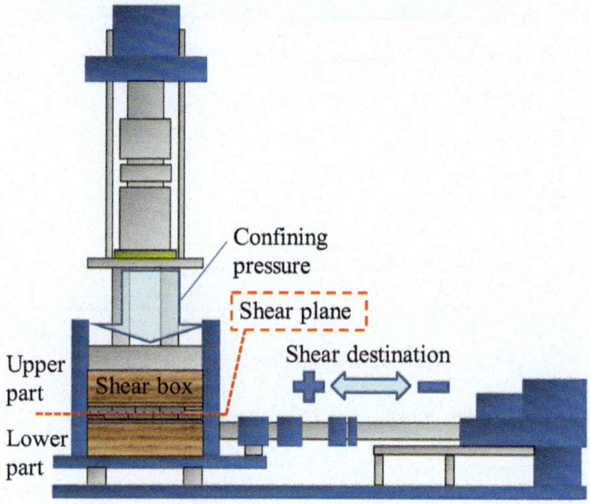

Fig. 10.5 Large size direct shear testing apparatus

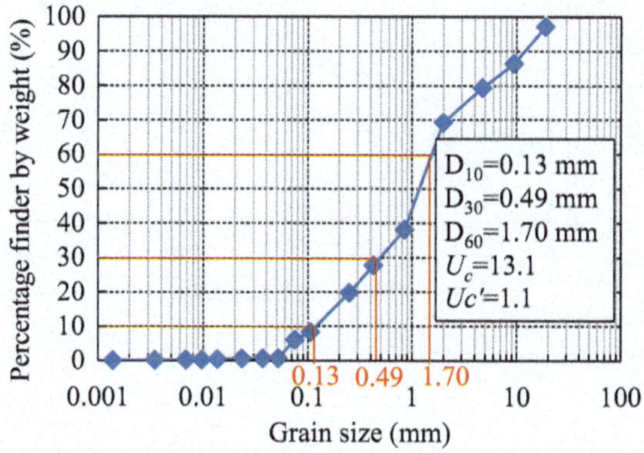

Fig. 10.6 Grain size distribution curve of the filled soils in tires

that was collected from the site. Density of the filling soil was determined based on the normal distribution of the wet density. Since water content at the site varies greatly depending on the season, a uniform value of 20% was used for simplicity.

Direct shear test under static loading

Static direct shear test was first performed to estimate the coefficient of friction between the tires. During the direct shear test, the area of contact between the tires (hereinafter referred to as contact area) changes with the increase of horizontal displacement (Fig. 10.7). Therefore, the normal stress-shear stress relationship was obtained by converting into the stress that only the tire surface purely bears.

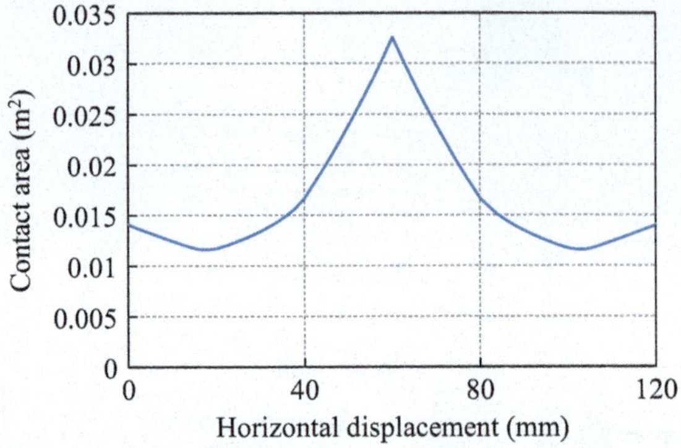

Fig. 10.7 Relationship between contact area and horizontal displacement

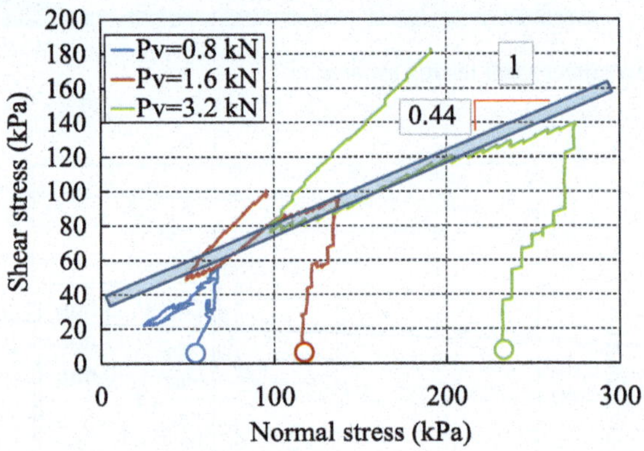

Fig. 10.8 Stress path after area correction

Figure 10.8 shows the stress path obtained from the experiments at three different values of the vertical load (Pv). The vertical load shown in the figure was obtained by multiplying the vertical stress with the cross-sectional area of the shear box (0.08 m^2). The blue band in the figure was created so that the red and green legends in Fig. 10.8 cover the range showing the approximate values. The coefficient of friction between tires was obtained to be 0.44, based on the slope of the curve.

Direct shear test under dynamic loading

To examine the damping characteristics of tires, cyclic direct shear test was performed under different normal loads. As in the static direct shear test, since the shear stress that the tire bears changes with the horizontal displacement, the relationship between normal stress and shear stress was obtained using the necessary area correction, as explained before.

During the cyclic loading, the stress control method was adopted, in which the shear direction is reversed when the set shear stress value is reached. During the

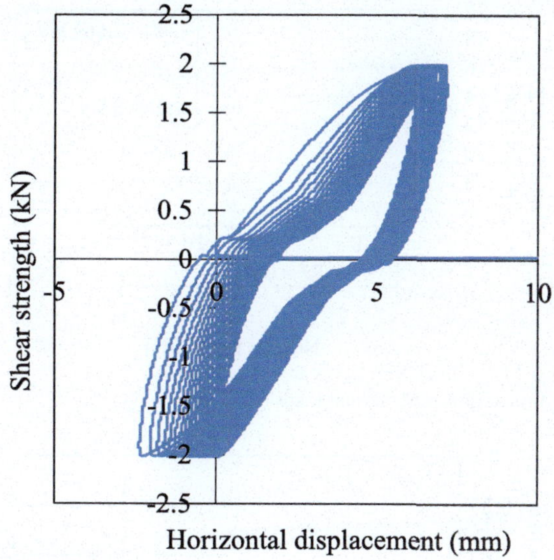

Fig. 10.9 An example of hysteresis loop

tests, the shear direction was set to reverse, when the shear resistance reached 2.0 kN. Assuming the period of cyclic load to be a long-period vibration wave, a sine wave with a frequency of 0.01 Hz was adopted. The number of cycles was set to 20 (the same value used, in general, for predicting the behavior of soils), at which the any geomaterials come to the stable state. A hysteresis loop, thus, could be drawn for each cycle within the horizontal displacements in which the designated shear force is mobilized. By calculating the area within the hysteresis loop for each cycle, the energy lost due to the repeated friction between the tires can be obtained. Finally, the damping constant was obtained from the energy loss and the shape of the hysteresis loop.

Figure 10.9 shows the experimental results for the normal load Pv = 4 kN. It can be confirmed that unlike materials such as soils, the shape of the hysteresis loop is asymmetrical with axis of symmetry moving away from the origin. Also, as described before, the strength development between tires is greatly affected by the contact area. The energy loss and damping constant were calculated for each cycle from the hysteresis loop shown in Figs. 10.9. Absorbed energy and damping constants for each cycle are plotted against each normal load as shown in Figs. 10.10 and 10.11, respectively. From the two figures, it can be confirmed that the smaller the normal load with respect to the shearing force, the larger the energy loss and damping constant. This suggests that the tire with the smaller normal load (smaller surcharge) on the upper part of the backfill soil has a greater damping effect on the seismic motion.

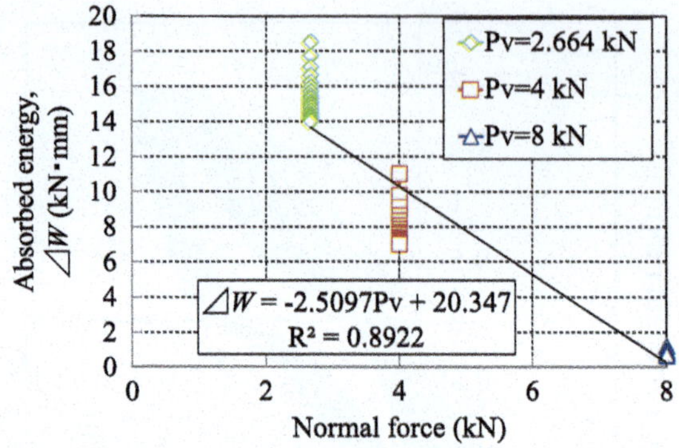

Fig. 10.10 Energy loss at different vertical loads

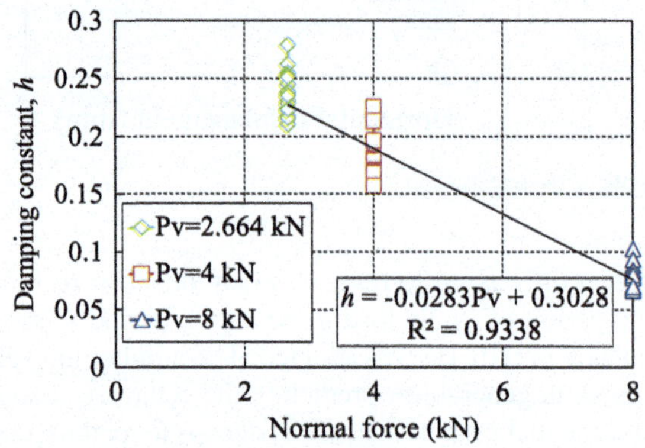

Fig. 10.11 Damping constant at different vertical loads

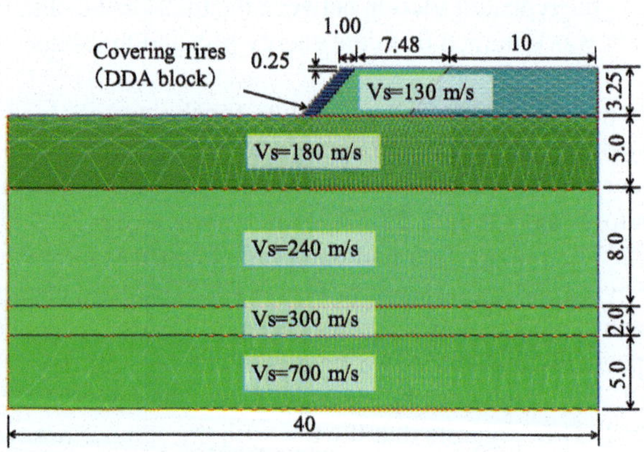

Fig. 10.12 Numerical model of embankment (unit in m)

10.2.3 Numerical Investigation on Performance Evaluation of an Embankment Under Earthquake Loading

Numerical simulation of an embankment with and without tire protected slope was performed using the frictional characteristics and damping characteristics obtained from the laboratory tests described above. In the numerical simulation, Manifold Method (MM) was adopted, which is a combination of Finite Element Method (FEM) and Discontinuous Deformation Analysis (DDA) [1, 14, 15]. This kind of hybrid numerical method was considered to be the most suitable for analyzing the retaining wall, where continuum (soils) and discontinuum (tires) must be dealt with simultaneously. In this study, the tire retaining wall was modeled as discontinuous block in the DDA, while the soils were modeled using FEM.

Analysis condition

Based on the results of the PDCP test and surface wave exploration at the sites, as described before, a numerical model of soil structure, including the much deeper soil layer, was constructed. The backfill soil of the tire retaining wall was modeled as a non-linear material. The safety of the embankment during an earthquake was examined for two cases: Conventional slope without any retaining wall and the presence of tire cover in the slope in the form of a retaining wall.

Figure 10.12 shows the numerical model together with the shear wave velocity Vs, which were obtained from the field survey. The input seismic motion is applied at the bottom of the model shown in Fig. 10.13. This is the waveform that was recorded at KiK-net station of Rikuzentakata city (which is the closest to the tire-covered embankment structure) during the 2011 off the Pacific coast of Tohoku Earthquake. The waveform was applied in the analyses after angle correction, in the direction of slope of embankment. The damping constants in each soil layer were calculated using the results in Fig. 10.11.

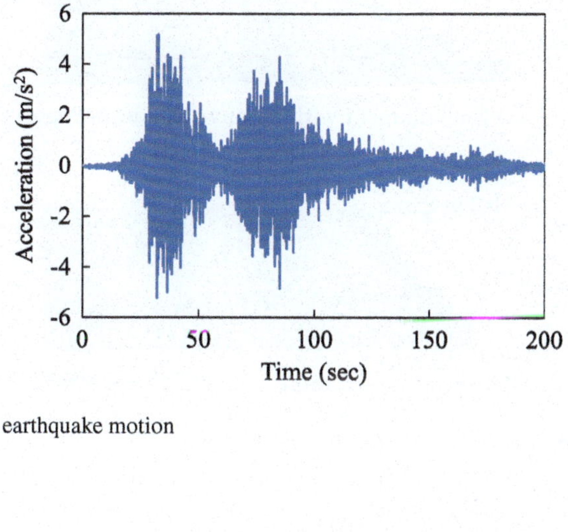

Fig. 10.13 Input earthquake motion

Analysis results

Figure 10.14 shows the shear strains developed within the embankment at the end of the seismic loading. The strain shown here is magnified by a factor of 2 to enhance the visibility. Similarly, Fig. 10.15 shows the deformation (magnified by a factor of 2 to enhance the visibility) of the embankment at the end of the seismic loading. In both the figures, the differences in stability of the embankment with the presence and absence of tire covering can be clearly observed.

Focusing on Fig. 10.14, it can be seen that, in the case of embankment having no covering tires as reinforcement (Fig. 10.14a), shear strain develops on an arc form, starting from the toe to the top of the embankment at an early stage after excitation,

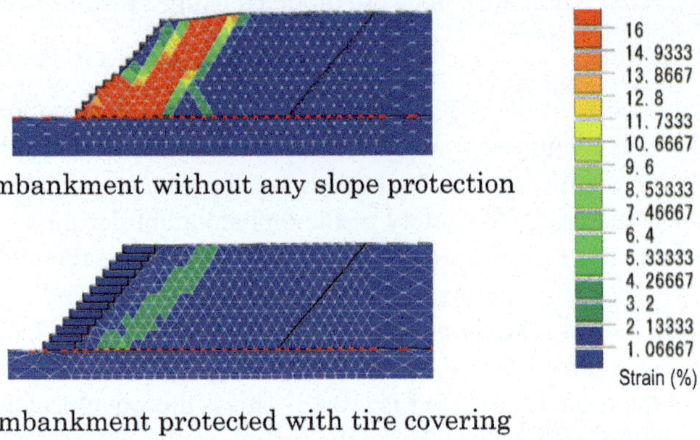

Fig. 10.14 Distribution of shear strain at the end of excitation

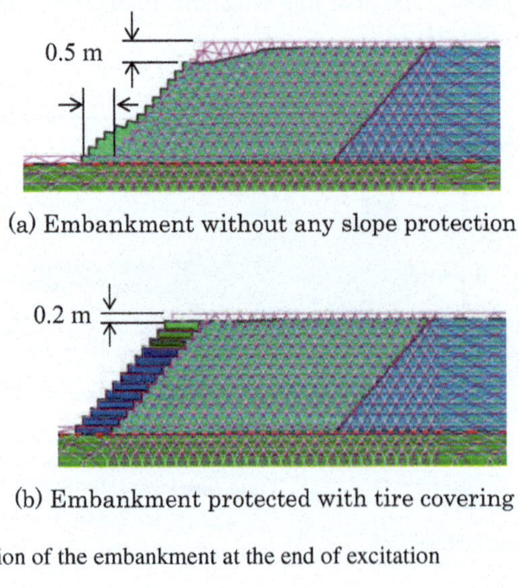

Fig. 10.15 Deformation of the embankment at the end of excitation

and a maximum strain of more than 15% developed at the end of the excitation. On the other hand, in the case when the embankment was covered with tires in the form of a retaining wall (Fig. 10.14b), the development of shear strain could be significantly suppressed with less than 5% at the maximum.

Focusing on Fig. 10.15, it can be confirmed that in the unreinforced embankment (Fig. 10.15a), the horizontal displacement increases with the development of shear strain, and the embankment undergoes significant deformation at the end of the excitation. However, the same embankment, while reinforced with tires (Fig. 10.15b), the deformation is minimal. From these results, it can be said that the reinforcement of the embankment with tire covering suppresses the shear strain and the horizontal displacement, and thus, could maintain the stability of the embankment during an earthquake.

The response accelerations were calculated at two points (one on the top of the tires and other 2 m away on the top of the backfill) as shown in Fig. 10.16. The calculated accelerations are shown in Figs. 10.17 and Fig. 10.18 for the two cases analyzed. Comparing the two figures, it can be seen that, there is no significant difference in the

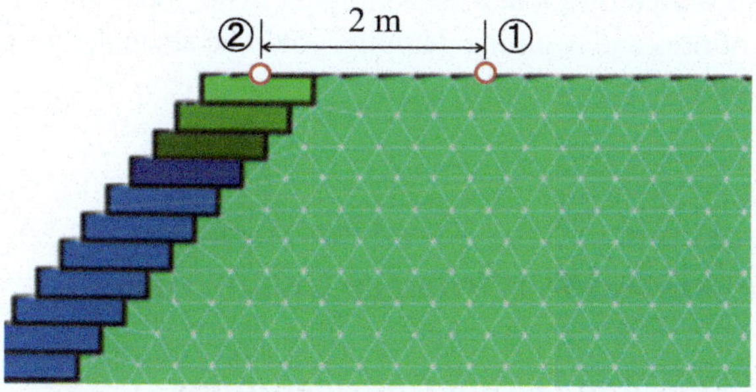

Fig. 10.16 Observation points of response acceleration

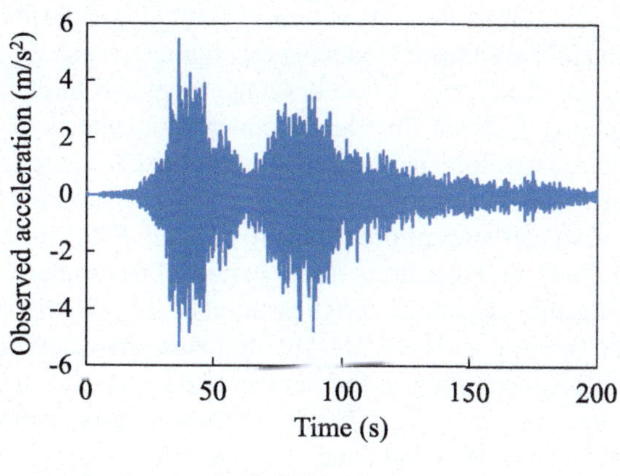

Fig. 10.17 Response acceleration at point① (Without protective measure)

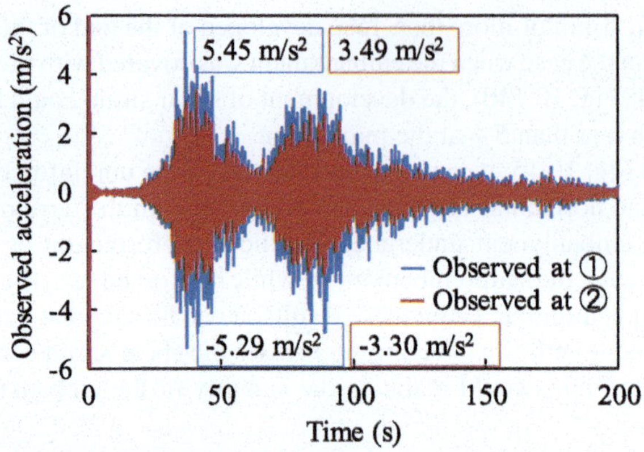

Fig. 10.18 Response acceleration at point② (With protective measure)

response acceleration on the top of the backfill (Location① of Fig. 10.16). However, it could be seen that the response acceleration at the top of the tires (Location② of Fig. 10.16) is decreased by about 35% compared to the top of the embankment. The viscoelasticity of tires and resulting damping could be attributed to such decrease in seismic acceleration.

10.3 Applications of Tire Shreds in Highway Embankments for Drainage Enhancement

10.3.1 Test Sites and Design Considerations

A full-scale field trial on the use of tire shreds as materials for drainage layer in a highway embankment, under the jurisdiction of East Nippon Expressway Co., Ltd., was undertaken in Hokkaido (near Houbetsu east tunnel), Japan. Figure 10.19 shows the embankment and the layout of the drainage layer. To implement the project, extensive permeability tests on tire shreds under simulated field conditions were required beforehand. Therefore, permeability tests were carried out using a prototype apparatus designed specifically for evaluating drainage potentiality by considering the compression characteristics of tire shreds to be used. Since tire shreds are highly compressible materials, and therefore, it was expected that under vertical load from the embankment, the drainage characteristics might be significantly altered according to the levels of compression. Therefore, the drainage characteristics of tire shreds need to be determined according to the degree of compression that the tire shreds will be subjected to. To evaluate the influence of these factors, compression tests and permeability tests were at first conducted on a laboratory scale.

Fig. 10.19 Embankment with tire shreds drainage layer

10.3.2 Laboratory Tests

Test materials

Tire shreds, composed mainly of non-spherical particles and sizes ranging from 4.75 to 19 mm, were used in the tests. The tire shreds contain the steel and textile cords embedded inside the rubber as in the original scrap tire. Particle density of the tire shreds was 1.20 g/cm^3. Bulk densities of the compacted specimen were ranging from 6.5 kN/m^3 to 8.5 kN/m^3 depending on the compaction energy.

Compression test

One-dimensional compression tests were carried out in laboratory to evaluate the compressive strains of tire shreds under incremental vertical loading. A large-sized test apparatus comprised of a specimen mold 300 mm in diameter and 400 mm in height was used. A piston head of size 297 mm in diameter was used for loading the specimen in the mold. All specimens were compacted inside the mold, and the mold was extended and locked in its most lengthened state ($h = 400$ mm) before setting it for the compaction. After compaction, the mold was turned upside down, peripherals were attached, and sliding mechanism was unlocked for allowing the compression to take place during the testing.

Compressive strains of the specimens obtained at different vertical stress are shown in Fig. 10.20 for two different initial bulk densities of tire shreds. The specimen with higher initial bulk density (6.5 kN/m^3) experiences high level of compressive strain under the vertical loading, in spite of the fact that both were heavily compacted

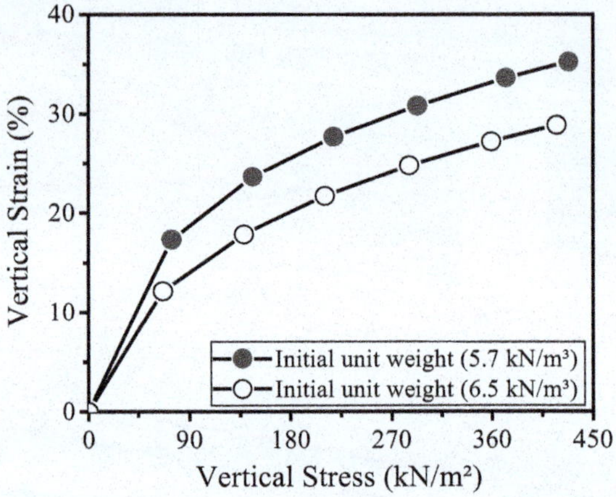

Fig. 10.20 Compressive strain of tire shreds specimens (Modified from [10])

inside the mold. Tire shreds, made up mainly of rubber material, are susceptible to elastic deformation during compaction, an inherent characteristic of such material. Therefore, it tends to rebound to its original state when the load is released. Consequently, it can be presumed that high void ratio can be sustained by tire shreds even after experiencing very heavy compaction.

Permeability test

Permeability tests on soil at the site (SCJV), conventionally used coarse gravel (Grvl) and newly proposed tire shreds (SG) samples (Fig. 10.21) were conducted in laboratory using a newly developed large-sized test apparatus (shown in Fig. 10.22). The apparatus is comprised of a specimen mold designed to work in the vertical water-flow condition. To evaluate the drainage behavior of tire shreds at different compressive loads, the length of the cylindrical mold was maintained in such a way that it could be mechanically adjusted. Therefore, tire shreds specimen inside the mold could either get compressed or expanded upon shortening and lengthening of the mold. Within the mechanical limits of this apparatus and/or the types of samples used, permeability tests could be performed on the tire shreds specimen under different compressive loading.

Fig. 10.21 Materials used for testing

Fig. 10.22 Test apparatus [10]

The permeability coefficients of the three different materials in Fig. 10.21 are shown in Fig. 10.23. It can be seen from the figure that the soil specimens collected from the site, which were compacted to different densities, possess quite different permeability coefficients. Densely compacted soil, SCJV(J2), with density of 18.0 kN/m^3 shows permeability coefficient of the order 10^{-7}–10^{-6} cm/s, which indicates that the field soil has very low permeability, and thus additional measures must be taken to maintain stability of the structure to be constructed using such soil. Conventionally used coarse gravel (Grvl) material shows permeability of the order 10^{-1}–10^{2} cm/s. Permeability coefficient of tire shreds was found to be in the higher-order ($\times 10^2$ cm/s), which is comparable to conventionally used highly permeable geomaterials (e.g., gravel). It is worthwhile mentioning here that there is a limitation on the maximum rate of water flow in the large-scale test apparatus. Therefore, the permeability coefficients obtained are expected to be a little conservative. Also, Darcy's law, which is valid for a wide range of soil types and hydraulic gradients, is

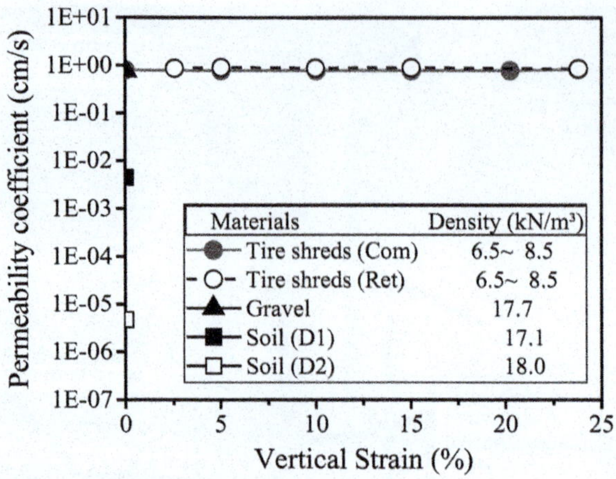

Fig. 10.23 Permeability coefficients of different geomaterials (Modified from [10])

assumed to be valid in this test too. Comparative evaluations on materials using the same set-up were performed here for qualitative permeability characteristics, which show no adverse effect on permeability due to such laminar flow assumption.

During the testing, tire shreds specimen (SG) was compressed (referred to here as Com) and then allowed to return to its original state (referred here as Ret) gradually in the mold. Figure 10.23 shows that the permeability coefficients corresponding to various compressions (Com) and also returned (Ret) states remain almost unchanged. This clearly indicates that tire shreds maintain almost equal levels of permeability coefficients to that of gravel even after undergoing compression as high as 23–24% (equivalent to the load of about 15 m high soil layer above the tire shreds drainage layer).

10.3.3 Field Trial

The fact that permeability of the tire shreds does not decrease despite a high level of compression that might be induced in the field due to soil surcharge, has paved the way for their utilization in drainage enhancement. High permeability characteristics, together with non-clogging nature of tire shreds [11], had encouraged us to undertake a full-scale field trial at Hokkaido, Northern Japan on the use of tire shreds as drainage layer in a highway embankment. Cross-sectional view of the embankment, where tire shreds drainage layers were installed, is shown in Fig. 10.24. Design thickness of the blanket type drainage layer was 400 mm each with an inclination of 5% toward the embankment face. Thickness of drainage layers was arrived at by considering the relevant change of thickness of tire shreds due to compression. The design length and width of the tire shreds drainage layers were 40 m and 10 m, respectively. The target

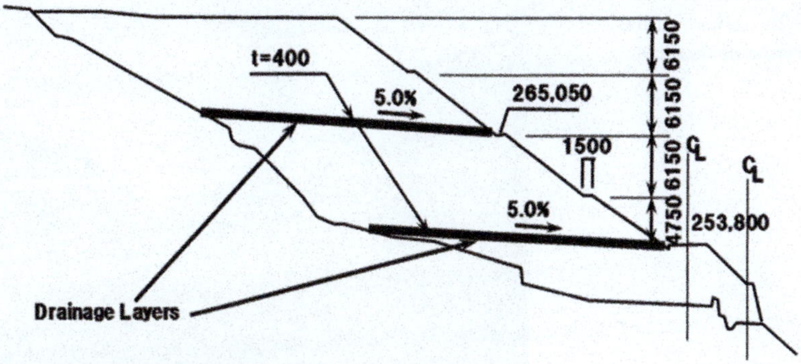

Fig. 10.24 Line diagram of the embankment

coefficient of permeability was of the order 10–1 ~ 100 cm/sec, which is equivalent to the coefficient of permeability of gravel that was originally planned to be used in this field trial.

A total of 240 tons (400 m^3 in volume) of tire shreds were used in the construction of the drainage layers. Conventional equipment and machineries were used in the construction of this embankment. Construction procedures involving the laying of tire shreds drainage layer (dumping and compacting) are shown in Fig. 10.25. The completed embankment after one year is shown in Fig. 10.26. It is worthwhile mentioning here that the *in-situ* consolidations of tire shreds drainage layers were measured for about a year since their placement in the embankment [10, 11], and the results are found well in parity with the results of the consolidation tests conducted in laboratory. The degree of *in-situ* consolidation at the bottom tire shreds layer was found to be 27%.

(a) Dumping of tire shreds (b) Compacting the tire shreds layer

Fig. 10.25 Construction with tire shreds

Fig. 10.26 A view of the completed embankment

10.4 Concluding Remarks

Two case studies, which are presented here, paid particular attention to the drainage, compressibility, vibration absorption, and ductility characteristics of scrap tire-derived geomaterials. The following are some of the main conclusions.

1. Damage to the tire retaining wall could be prevented during the earthquake due to confining effect of each individual tire, friction between the tires, ductility, and damping inherent in tires. All these factors combined, reduced the earth pressure and water pressure acting on the wall during the earthquake and the tsunami, thus attributing to such good performance of the wall.
2. The presence of tire cover in an embankment suppresses the shear strain and horizontal displacement of the embankment, thus, ensuring the stability of an embankment during earthquake.
3. Tire shreds can maintain the highly permeable nature of the material even under the action of high compressive load such as in an embankment, and thus the use of tire shreds as drainage enhancing layer was validated through a field trail. The performance of tire shreds drainage layer in the embankment has been found to be very satisfactory till date.

Scrap tire recycling as geomaterials can lead to sustainable design that comes with extra reduction in construction cost of a project. It can also lead to a cost-effective design and retrofitting of geotechnical structures against earthquake-induced damage. To realize the Sustainable Development Goals (SDGs), use of such materials needs to be encouraged as the future materials in civil engineering projects.

References

1. Chen, G.: A view on manifold method comparing with finite element method. In: Computational Mechanics. Springer, Berlin, Heidelberg (2007). https://doi.org/10.1007/978-3-540-75999-7_188
2. Hazarika, H.: Structural stability and flexibility during earthquakes using tyres (SAFETY). In: Hazarika, Yasuhara (eds.) A Novel Application for Seismic Disaster Mitigation. Scrap Tire Derived Geomaterials-Opportunities and Challenges, Taylor and Francis, London, pp. 115–125 (2007)
3. Hazarika, H., Fukumoto, Y.: Sustainable solution for seawall protection against tsunami-induced damage. Int. J. Geomech. 16(5) (2016). https://doi.org/10.1061/(ASCE)GM.1943-5622.0000687
4. Hazarika, H., Kohama, E., Sugano, T.: Underwater shake table tests on waterfront structures protected with tire chips cushion. J. Geotech. Environ. Eng. ASCE, USA **134**(12), 1706–1719 (2008)
5. Hazarika, H., Okada, H., Hara, T., Ueno, M., Ohsumi, T., Yamanaka, M., Yamazaki, T., Kosaka, N, Minowa, H., Furuichi, H: Case studies of geotechnical damage by the 2011 off the Pacific Coast of Tohoku Earthquake and tsunami in Japan. In: Proceeding of the 15th World Conference on Earthquake Engineering, Lisbon, Portugal, Paper No 4796 (2012)
6. Hazarika, H., Pasha, S.M.K., Ishibashi, I., Yoshimoto, N., Kinoshita, T., Endo, S., Karmokar, A.K., Hitosugi, T.: Tire chips reinforced foundation as liquefaction countermeasure for residential buildings. Soils Found. **60**(2), 315–326 (2020)
7. Hazarika, H., Yasuhara, K.: Scrap Tire Derived Geomaterials - Opportunities and Challenges. Taylor & Francis, London, UK (2007)
8. Hazarika, H., Yasuhara, K., Kikuchi, Y., Karmokar, A.K., Mitarai, Y.: Multifaceted potentials of tire derived three dimensional geosynthetics in geotechnical application and their evaluation. Geotext. Geomembr. **28**(3), 303–315 (2010)
9. Karmokar, A.K.: Use of scrap tire derived shredded geomaterials in drainage application. In: Hazarika, Yasuhara (eds.) Scrap Tire Derived Geomaterials - Opportunities and Challenges. Taylor & Francis, London, pp. 127–138 (2007)
10. Karmokar, A.K., Takeichi, H., Kawasaki, H., Kawaida, M., Kato, Y., Mogi, H., Yasuhara, K.: Consolidation behavior of scrap tire shreds for their use as geomaterials. In: Proceedings of 42nd Annual Conferences of Japanese Geotechnical Society, Japan, pp. 675–676 (2007)
11. Karmokar, A.K., Takeichi H., Yasuhara K.: Triaxial compression test on granulated scrap tire rubber for their use as lightweight geomaterials. In: Proceedings of the International Workshop on Lightweight Geo-Materials, Tokyo, Japan, pp. 127–134 (2002)
12. Nakamura, Y.: A method for dynamic characteristics estimation of subsurface using micro tremor on the ground surface. Q. Rep. of RTRI **30**(1), 25–33 (1989)
13. Okada, K., Sugiyama, T., Noguchi, T., Muraishi, N.: A correlation of soil strength between sounding tests on embankment surface. Tsuchi to kiso, Japan. Geotech. Soc. **40**(4), 11–16 (1992). (in Japanese)
14. Shi, G.H.: Manifold method of material analysis. In: Transactions of the 9th Army Conferences on Allied Mathematics and Computing, Minneapolis, USA, pp. 57–76 (1991)
15. Shi G.H., Goodman, R.E.: Discontinuous deformation analysis. In: Proceedings of the 25th U.S. Symposium on Rock Mechanics, pp. 269–277 (1984)
16. Yasuhara, K.: Recent Japanese experiences on scrapped tires for geotechnical applications. In: Hazarika, Yasuhara (eds.) Keynote Paper, Proc. Int. Workshop on Scrap Tire Derived Geomaterials. Opportunities and Challenges. Taylor & Francis, London, pp. 17–40 (2007)
17. Yasuhara, K., Hazarika, H., Fukutake, K.: Applications of scrap tires in civil engineering. Doboku Gijutsu **61**(10), 79–86 (2006). (in Japanese)

References

[illegible]

Chapter 11
Log Piling Method for Liquefaction Mitigation and Carbon Stock in Large Residential Area, Chiba, Japan

Atsunori Numata and Katsuya Matsushita

11.1 Introduction and Background of Developing LP-LiC Method

Because global warming is a serious issue, the aim of construction work should be not only completing the construction work but also mitigating the emission of greenhouse gases. Indeed, it would be even better if the construction work helped mitigate global warming. To realize this concept, a method that uses logs to mitigate soil liquefaction and store carbon, called LP-LiC method has been developed [2]. In Japan, forest resources are currently extremely abundant and there are calls to harvest and use them. As such, a ground improvement method that uses a large amount of wood is highly relevant.

A. Numata (✉)
Tobishima Corporation, Tokyo, Japan
e-mail: atsunori_numata@tobishima.co.jp

K. Matsushita
Misawa Homes Institute of Research and Development Co. Ltd., Tokyo, Japan
e-mail: katsuya_matsushita@home.misawa.co.jp

H. Hazarika et al. (eds.), *Practices and Trends in Ground Improvement Techniques*, Developments in Geotechnical Engineering,
https://doi.org/10.1007/978-981-19-3322-6_11

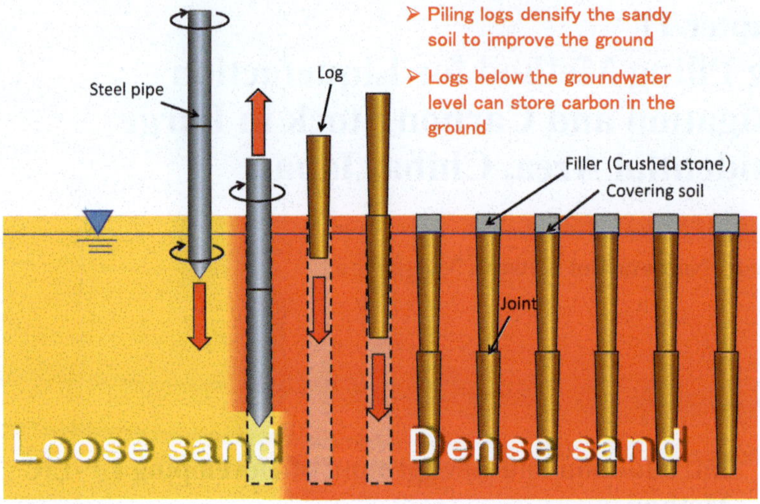

Fig. 11.1 Countermeasures principle and procedure of the LP-LiC method [2]

11.2 Overview of LP-LiC Method

Figure 11.1 [2] shows countermeasures principle and procedure of the LP-LiC method. First, a steel pipe whose tip is closed is pushed with rotation into the ground. Once the steel pipe has reached its intended depth, it is pulled out with rotation. Next, via the hole that was just made, a log is pushed into the ground until its top is below the groundwater level, whereupon the top of the log is covered with soil to prevent contact with the air. Two logs can be connected vertically with a steel joint. Finally, filler is inserted with vibration into the hole. Through this process, the logs densify the ground between them, thereby increasing their liquefaction resistance. In addition, because the logs do not deteriorate biologically in the water, the carbon that they have absorbed from the atmosphere via photosynthesis can be stored in the ground almost indefinitely.

11.3 Construction

The LP-LiC method was implemented as countermeasures against liquefaction for a large residential ground in Chiba Prefecture, Japan. Figure 11.2 [2] shows the construction site location. Figure 11.3 [2] shows construction situation with the LP-LiC method.

Liquefaction occurred on the west side of the construction site during the 2011 Great East Japan Earthquake. This site is coastal reclaimed land, and because of the high liquefaction potential at the site, it was decided to implement liquefaction

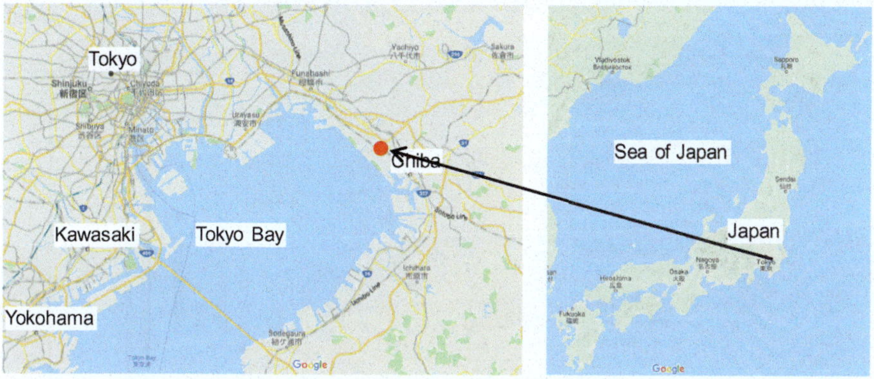

Fig. 11.2 Construction site location [2]

Fig. 11.3 Construction situation with the LP-LiC method [2]

countermeasures over the entire site. In addition, because the site is in a residential area, the construction method had to involve low vibration, low noise, no surplus soil, and no displacement of the surrounding area, in addition to having a positive impact on the environment. For this reason, LP-LiC was chosen for this site.

Figure 11.4 [2] shows a plan view of the site and the arrangement of logs there. In total, 13,420 logs with a diameter of 0.15 m and a length of 4 m were used. The

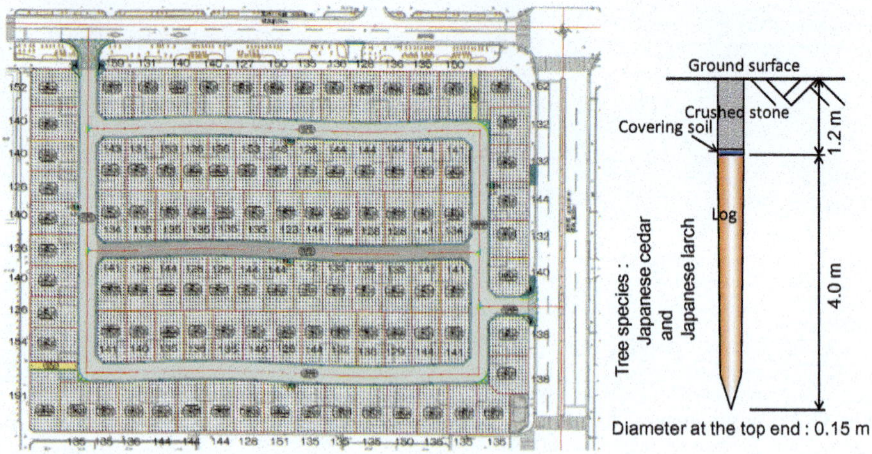

Fig. 11.4 Plan view of the site and the arrangement of logs [2]

depth of the top of each log was 1.2 m, and the distance between adjacent logs was 1 m. The logs came from trees of the species Cryptomeria japonica (Japanese cedar) and Larix kaempferi (Japanese larch). The target performance for liquefaction was that the safety factor for liquefaction to ground shallower than 5 m should be at least 1.

11.4 Design Against Liquefaction with LP-LiC Method

Five standard penetration tests were conducted at this site, and Fig. 11.5 [2] shows an example of a log being bored in one such test. Near the upper end of the As sand layer, there is a layer with low N-value that has the possibility of liquefaction. Table 11.1 gives the values of the minimum safety factor F_L against ground liquefaction for each boring shallower than 5 m. For borings 2 and 3, it is clear that $F_L < 1$, and the intention was to use LP-LiC to achieve $F_L > 1$ at these two locations. Table 11.1 also gives the values of the minimum safety factor against liquefaction designed by the aforementioned LP-LiC. In this calculation, with reference to the Architectural Institute of Japan's Recommendations for Design of Building Foundations [1] the earthquake magnitude was assumed to be 7.5 and the maximum horizontal acceleration at the ground surface was assumed to be 200 Gal. In this case, with reference to guidelines by the Ministry of Land, Infrastructure, Transport, and Tourism in Japan, the target depth for liquefaction measures was taken to be less than 5 m. The depth of the log tops was determined from the groundwater level and capillary height.

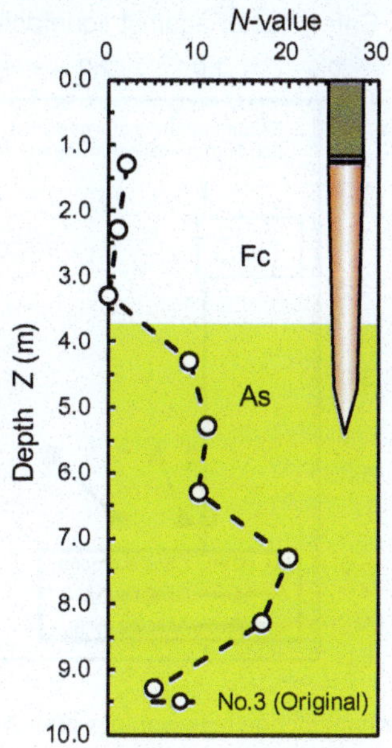

Fig. 11.5 Example of boring log by a standard penetration test [2]

Table 11.1 Minimum safety factor F_L against ground liquefaction for each boring shallower than 5 m [2]

Boring No.	Minimum safety factor F_L against liquefaction (from GL 0.0 m to GL − 5.0 m)	
	Original ground	Designed
1	2.46	2.46
2	0.96	1.17
3	0.97	1.18
4	2.40	2.43
5	1.16	1.44

11.5 Results of Liquefaction Countermeasures

Figure 11.6 [2] shows values of the safety factor against liquefaction between logs after log piling, those for the original ground, and the designed values. The values of the safety factor, which were below 1 before log piling, clearly increased and exceeded both 1 and the designed values after log piling. Similarly, the initial and designed values at boring 2 were 0.96 and 1.17, respectively, but the value rose to 1.63 after log piling.

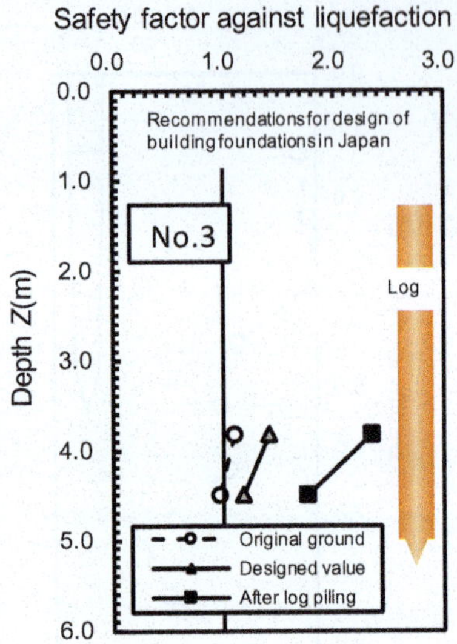

Fig. 11.6 Safety factors against liquefaction between logs after log piling [2]

11.6 Carbon Storage Effect

One of the major features of LP-LiC is that carbon fixed to trees by photosynthesis from the atmosphere can be stored in the ground almost indefinitely. The amount of carbon storage at this construction site by converting it to the equivalent amount of CO_2 was calculated. The amount S_{Log} of carbon stored in the logs is calculated by

$$S_{Log} = V_{Log} \times \rho_{Log} \times K_C \times K_{CO2C}, \tag{11.1}$$

where V_{Log} is the total volume of the logs (m^3), ρ_{Log} is the density of the logs (t/m^3), K_C is the mass ratio of carbon to logs (=0.5), and K_{CO2C} is the coefficient for converting from carbon to CO_2 (=44/12). The amount of CO_2 emitted by the construction was estimated by measuring the amount of fuel used.

Figure 11.7 [2] shows the balance of carbon fixed by the logs and CO_2 emitted by the construction. The amount of carbon stored in the logs is 790.6 t-CO_2 and the amount of CO_2 emitted by the construction was 37.7 t-CO_2. Clearly, the amount of carbon stored in the logs exceeds the amount of carbon emitted by the construction by more than a factor of 10. Therefore, by using the LP-LiC construction method, the amount of CO_2 removed from the atmosphere was approximately 750 t-CO_2.

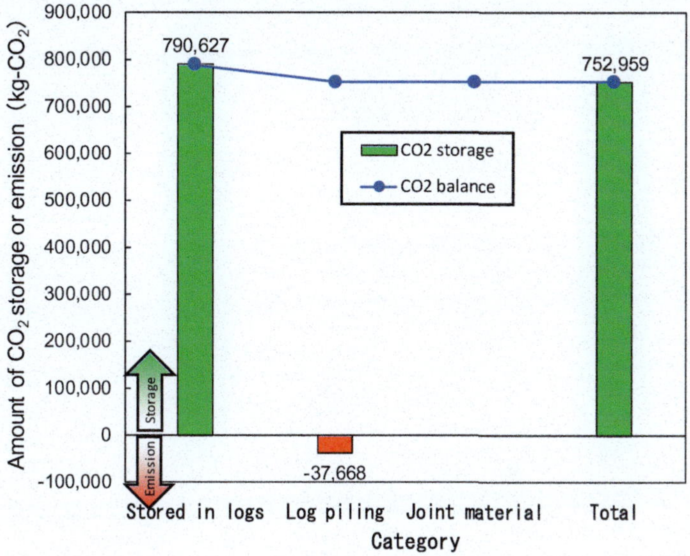

Fig. 11.7 Storage and emission amounts of CO_2 by construction [2]

11.7 Summary

An example of using LP-LiC as a countermeasure against liquefaction over a large residential area of 13,400 m^2 was introduced. As a result of using LP-LiC, the effect of the liquefaction countermeasure was found to exceed the design value. The carbon storage effect of LP-LiC was found to exceed the amount of CO_2 emitted by the construction by far more than a factor of 10. Therefore, this construction work contributed to a reduction in greenhouse gas.

References

1. Architectural Institute of Japan.: Recommendations for design of building foundations, pp. 61–72 (2001) (in Japanese)
2. Numata, A., et al.: Development of log piling method for liquefaction mitigation and carbon stock. Tobishima Tech. Rep. **63**, 13–21 (2014). (in Japanese)

Chapter 12
Ground Freezing for Shield Tunneling, Japan

Tsutomu Tsuchiya

12.1 Overview of the Projects

This chapter introduces two projects in which the ground freezing method was adopted. The first project is a railway construction project in Osaka. The Nakanoshima Subway Line, which opened in 2008, is a subway running on Nakanoshima between the Dojima River and the Tosabori River, and its total length is 2.9 km. Among the section, the shielding length of about 440 m (construction area No.4), which is between Oe-bashi Station and Naniwa-bashi Station, was located under the Mido-suji Subway Line, but there were many existing steel sheet piles, and they became an obstacle in the shield construction work of soil pressure shield machine. Since steel sheet piles were right underneath the existing subway tunnel structure, it was impossible to remove them from the ground, hence the ground freezing method was adopted.

In February 2019, a first operation using LNG fuel for the fire thermal power plant in Hokkaido has been started. For the construction of this thermal power station, a ground freezing method using CO_2 as a refrigerant was adopted to cut off the water at arrival point of the shield in the discharge tunnel—This is the second project introduced in this section. The drainage tunnel was a shielded tunnel with an inner diameter of 4.7 m connecting the starting shaft of the revetment and the drainage pipe buried 1 km off the coast and was constructed by a hydraulic shield under the sea bed on which 0.4 MPa water was pressured. Since the arrival point of the discharge tunnel was under the sea bed, it was difficult to install general ground improvement for shield arrival protection from the sea, and therefore, ground freezing method that allows construction within the shield tunnel was adopted. This project was the first to be implemented with the CO_2 system, which is a newly developed method that has

T. Tsuchiya (✉)
Chemical Grouting Co., Ltd., Tokyo, Japan
e-mail: t-tsuchiya@chemicalgrout.co.jp

H. Hazarika et al. (eds.), *Practices and Trends in Ground Improvement Techniques*, Developments in Geotechnical Engineering,
https://doi.org/10.1007/978-981-19-3322-6_12

various advantages, such as low environmental loads and could shorten construction period, compared with conventional methods.

12.2 Outline of Ground Freezing Method

The ground freezing method is a ground improvement technique in which groundwater (pore water) is frozen by using refrigerant. The ground freezing method is mainly used as a temporary work of large-scale shield construction in great depth because it can reliably cut off water. In addition, materials such as cement will not be injected into the ground, thus the technique has low environmental load.

The ground freezing method freezes the soil using two types of refrigerants. The secondary refrigerant is cooled by a freezer using the primary refrigerant, and the secondary refrigerant is circulated into pipes (freezing pipes) buried in the ground, and the ground around the pipe is frozen. The conventional ground freezing method has widely used CFCs (HCFC) or CFCs alternative (HFCs) as the primary refrigerant, and brine (antifreeze), as the secondary refrigerant. These are which is called a brine method, and the thermal difference between feeding and returning brine (sensible heat) will deprive the ground heat so that a frozen soil was made. Heat removed from the ground in the refrigeration cycle using CFCs (primary refrigerant) is radiated from the cooling tower to the atmosphere. Schematic diagram is shown in Fig. 12.1.

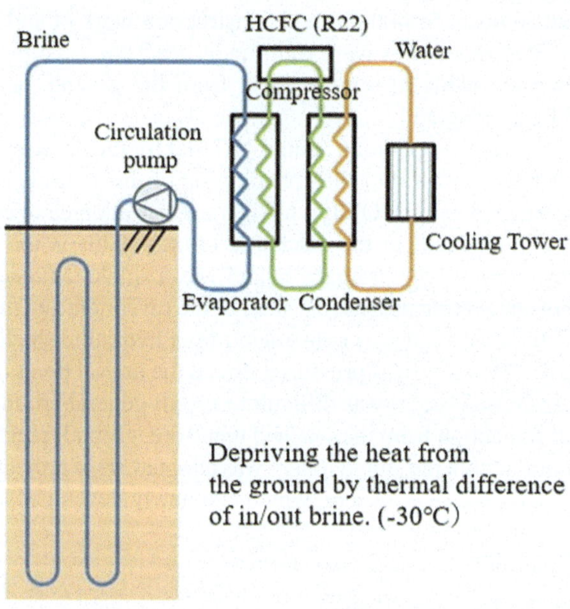

Fig. 12.1 Flow chart of Brine method [1]

However, from the viewpoint of global warming, the use of CFCs tends to be restricted worldwide. In Japan, HCFC production will be discontinued in 2020 due to CFC regulations based on the Ozone-Layer Protection Law, and HFCs are expected to be also regulated in the future. Therefore, development of a method using natural refrigerant with low global warming potential (Table 12.1) was required, and a ground freezing method using NH_3 as the primary refrigerant and CO_2 as the secondary refrigerant was developed in the late 2010's in Japan.

Figure 12.2 shows an illustrated schematic diagram of the CO_2 system. In the CO_2 system, CO_2 is circulated in the freezing pipe. Liquefied CO_2 is circulated in the freezing pipe, a part of which is vaporized by the heat of the ground, and returned to the refrigerator in a gas–liquid mixed phase. In other words, in addition to the sensible heat, CO_2 deprives the heat from ground by latent heat of the vaporization of CO_2 so that the frozen soil is created. In the refrigerator, the heat deprived from

Table 12.1 List of Global Warming Potentials (GWP)

Classification	Representative	GWP	Restriction year
CFC(CFC)	R-12	10,900	2009
HCFC(CFC)	R-22	1810	2020
HFC(altanative CFC)	R-404A	3920	Under investigation
Natural refrigerant (NH_3)	R-717	<1	
Natural Refrigerant (CO_2)	R-744	1	

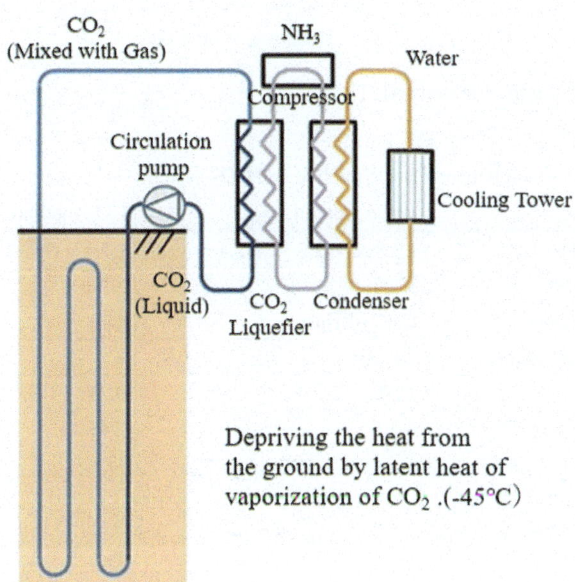

Fig. 12.2 Flow chart of CO_2 system [1]

ground is transferred to cooling water with circulating system of NH_3 which is a primary refrigerant and is released from the cooling tower to the atmosphere. Since the CO_2 system utilizes not only the thermal difference of the sensible heat but also the latent heat, the heat exchange rate per unit flow rate increases, and so CO_2 is able to can create the same volume of frozen soil with the flow rate per pipe being about 1/10 of that of the brine method. CO_2 is solid-to-gas (dry ice) at atmospheric pressure, but by applying pressure and maintaining it at 0.7–0.9 MPa, a phase change between gas and liquid is enabled (Fig. 12.3). Comparison between the items of the brine method and the CO_2 method is shown in Table 12.2.

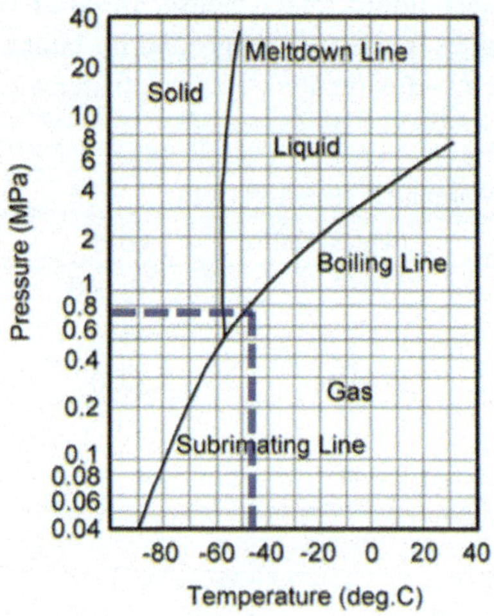

Fig. 12.3 Growth curve of frozen soil [1]

Table 12.2 Comparison between Brine method and CO_2 system

	Brine method	CO_2 method
Refrigerator size	L3.8 × W2.2 × H2.4 m	L7.9 × W1.2 × H2.0 m
Primary refrigerant	R-22	NH_3
Secondary refrigerant	$CaCl_2$ Brine	CO_2
Flow rate per pipe (l/min)	20–20	2
Viscosity (mPa/s)	14.71	0.21
Freezing temperature (°C)	−30	−45

12.3 Quality Control and Quality Assurance

The quality control/assurance of the ground freezing method is simpler and easier than that of general ground improvement. The quality of the ground freezing method can be managed only with real-time temperature measurement using temperature measuring pipes, and there is no need to take cores. During the freezing operation, a temperature below 0 °C indicates that the ground pore water is frozen, also indicating that frozen soil is being built/maintained. If the freezing operation is suspended, or forcible defrosting also indicates that pore water has been defrosted when the temperature exceeds 0 °C.

However, sufficient attention must be paid to frost heaving in quality control/assurance. Frost heaving is a phenomenon generated in accordance with the soil quality, groundwater, and applied loads. When applying the ground freezing method, it is necessary to suppress damage of the frost heaving by for example adjusting the thickness of frozen soil, in accordance with the site conditions.

12.4 Construction Challenges

In the following, two projects in which the ground freezing method was applied are introduced. One is the case in which ground freezing for the protection work of the tunnel underneath the existing structure, and the other is the case in which the ground freezing using newly developed CO_2 system is initially adopted as water cut off for the arrival point.

12.4.1 Case of Nakanoshima Subway Line in Osaka

In construction area No. 4 (Fig. 12.4) of Nakanoshima Line construction project, existing steel sheet piles at the intersection with Mido-suji Line were found. Plan view of the shield route, the structure of Mido-suji Line, and existing steel sheet piles are illustrated in Fig. 12.5. Though these steel sheet piles became obstacles to the shield excavation, they could not be removed from the ground because they were underneath the existing subway tunnel structure, hence they were removed by the vertical shaft and the horizontal shaft (Fig. 12.6). It was necessary to improve the ground as protection for excavation of the horizontal shaft, therefore in this project, the ground freezing method was adopted in consideration of safety, workability, and effects on the structure.

The ground freezing was implemented by the brine method. Arrangement of the freezing pipes is shown in Fig. 12.7. The target to be frozen was gravel layer and lies approximately 15 m deep. The determined frozen soil excavation range is based on the location of existing steel sheet piles shown in Fig. 12.8. As for the strength of the

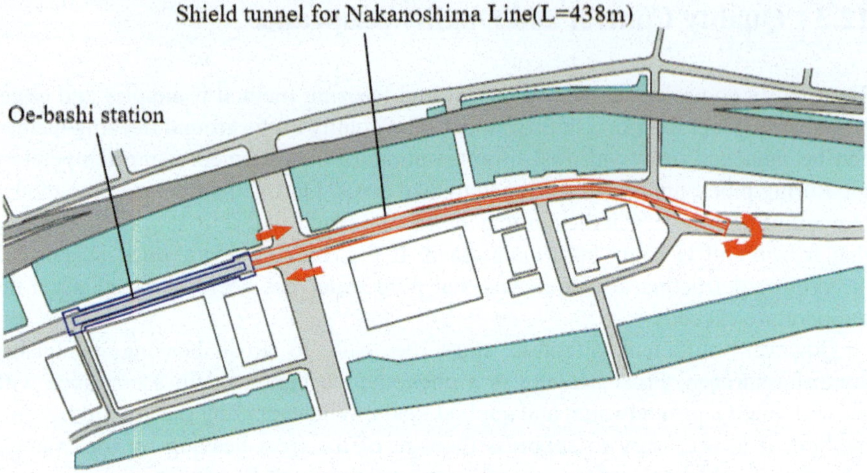

Fig. 12.4 Plan view of Nakanoshima Subway Line, construction area No. 4 [2]

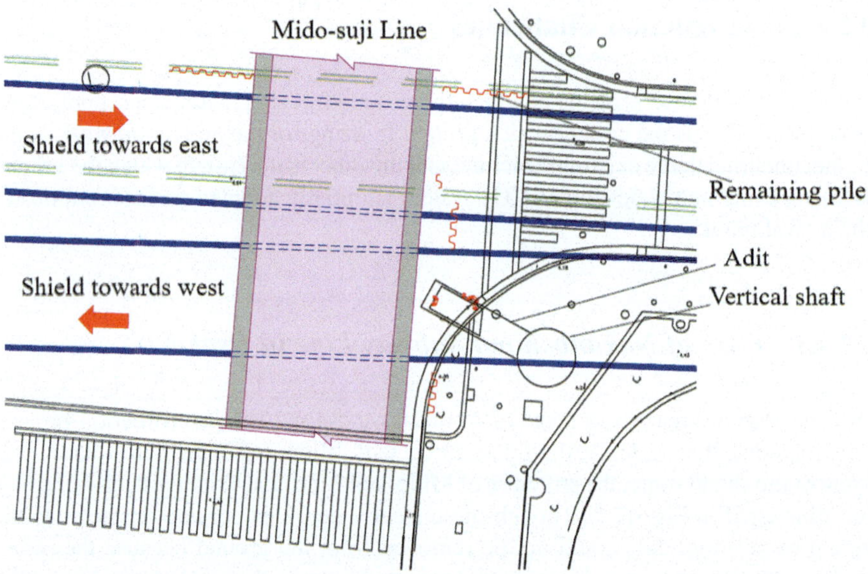

Fig. 12.5 Plan view of shield route—intersection area of Mido-suji line [2]

frozen soil, the average values shown in Table 12.3 were set as the design standard strength. The allowable stress was calculated using sandy soil, designed frozen soil temperature of −10 °C, and safety factor of 2.0, furthermore considering strength degradation due to the salinity of groundwater, and in conclusion, it was decided as shown in Table 12.4. Calculation was implemented again based on the allowable stresses, and the freezing area of the frozen soil was determined to be 1.5 m.

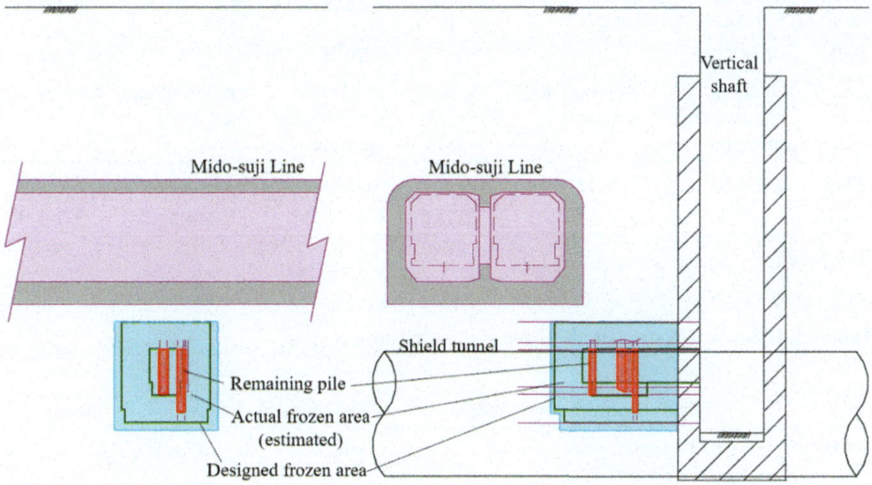

Fig. 12.6 Cross-sectional view of removal plan [2]

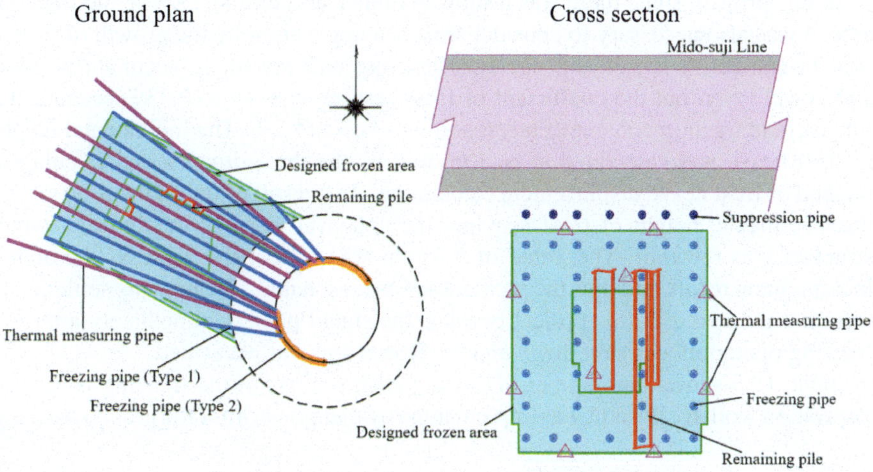

Fig. 12.7 Freezing pipe layout [2]

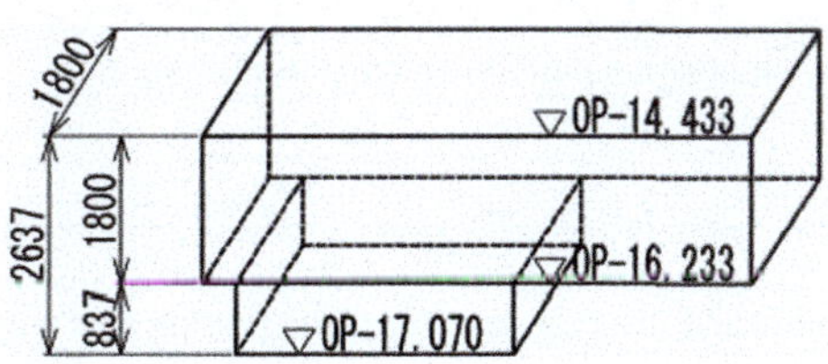

Fig. 12.8 Excavation area in frozen soil [2]

Table 12.3 Frozen soil strength

Soil	−10 °C			−20 °C		
	Uniaxial compressive strength	Bending and tension	Shear	Uniaxial compressive strength	Bending and tension	Shear
Clay	2.0–4.0	1.5–2.0	1.5–2.0	4.5–6.0	2.0–3.0	2.0–3.0
Sand	4.0–7.0	2.0–3.0	2.0–3.5	6.0–10.0	3.0–4.5	3.0–5.0

Unit MN/m²

Table 12.4 Allowable strength of frozen soil in this project

	Sand(Fs-2.0)		
	Uniaxial compressive strength	Bending and tension	Shear
−10 °C	853	388	425

Unit kN/m²

In this project, since there was a subway tunnel structure just above the freezing area, it was also necessary to consider frost heaving caused by the growth of frozen soil. In general, it is said that the frost heaving will not likely occur in the sand and gravel layer, but the coefficient of frost heaving was set to 0.33% because the fine particle fraction content of target soil was Fc = 11.2%. The incremental height of frost heave was calculated to be 1.66 mm by the three-dimensional calculation method of frost heave displacement calculation. As a result of the analysis by giving the incremental height of frost heaving, the upheaval value of the subway tunnel structure was 0.6 mm. The value of 0.6 mm was almost the same as the actual measurement result, and thus the prediction was reasonable. The thawing settlements were not measured in this project because the shield passed immediately after the freezing operation was shut down and the frozen soil was excavated.

Table 12.5 shows the number of days of each process of the freezing work. All excavation work for the adit was implemented by manpower after the ground freezing

Table 12.5 Total construct days for every constructing process

Working contents	Total day (days)
Installing freezing pipe, suppression pipe, and thermal measuring pipe	86
Freezing operation	42
Excavating adit	16
Removing sheet pile	4
Removing deep foundation and freezing pipe (for shield passing area only)	15
Total	163

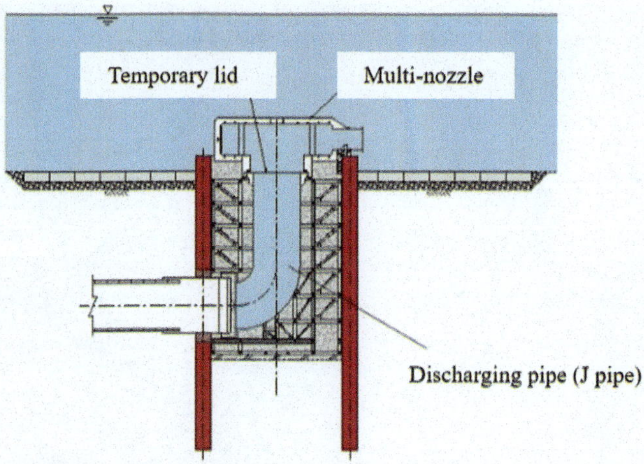

Fig. 12.9 Arrival point of drainage tunnel [3]

was completed. The temperature in the vertical shaft was −5 °C and the working environment was treated as cold district. The remaining steel sheet piles were removed, and the shield machine was able to pass beneath the subway tunnel structure safely.

12.4.2 Case of Discharge Tunnel of Thermal Power Station in Hokkaido

The ground freezing of CO_2 system was initially applied in the shield arrival protection project of the discharge tunnel of the thermal power station in Hokkaido. Figure 12.9 shows the structure of the arrival point of the discharge tunnel, which is called a J pipe because of its shape. The J pipe was filled with mortar and the clearance between the cutting surface and the shield was frozen to prevent water inflow. Since the J pipe is placed in the sea bottom, it is difficult to implement the construction by general ground improvement, and therefore, the ground freezing from the shield pit was adopted. At this time, the ground freezing of CO_2 system which has various advantages compared with the conventional method was adopted.

Freezing facility of this project is shown in Fig. 12.10, and the layout is shown in Fig. 12.11. The freezing facility is composed of a CO_2 reserve tank, a refrigerator, a dry cooler, and an NH_3 exclusion machine. In the conventional ground freezing, which is brine method, the freezing facility required large area, and it was necessary to install the equipment at a position as close as possible to the freezing pipe. On the other hand, since the freezing facility of the CO_2 system is more compact than brine system (width with only 1.2 m), the frozen soil can be created in parallel with the dismantling work of the shielding machine. Furthermore, as CO_2 system has allowed long-distance pumping, which enables to install pipes from afar, freezing facility

Fig. 12.10 Freezing facility [3]

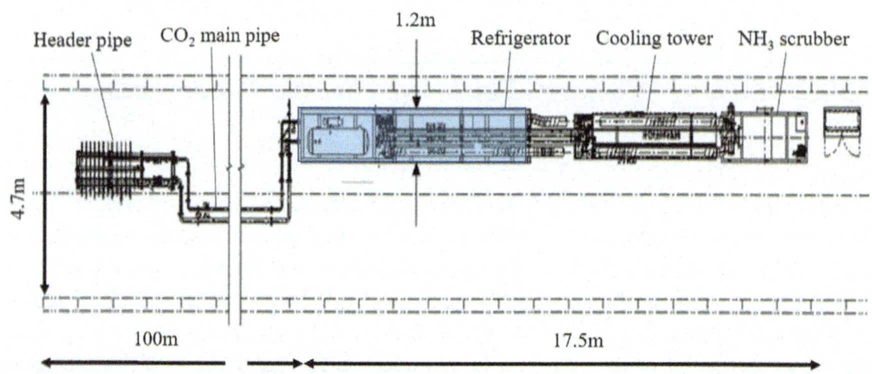

Fig. 12.11 Freezing facility layout [3]

could be installed behind the subsequent shield machine (approximately 100 m from the face). Utilizing the small diameter of the pipes, space for carrying out the disassembled materials was kept so that the ground freezing operation was performed without suspending the dismantling operation of the shield machine. During winter in Hokkaido, it was impossible to work in the outdoor, causing problem that were only 28 months of operation period left among the entire 54 months of the project, but this problem was solved by shortening the construction period with the CO_2 ground freezing method.

In this project, frozen soil was created by installing freezing pipes on the shield machine side (inside) and J pipe side (outside) and sandwiching them with freezing pipes (Fig. 12.12). In the freezing operation, the CO_2 was circulated at about 3 to 5L per minute per a freezing pipe. A thermos pipe was installed in the hood part of the J pipe, and the temperature was measured in real time. Water seepage was carried out in the freezing operation for 7 days from the start of freezing from the shield machine side. The thickness of the frozen soil was 485–750 mm, twentieth days after the start of the freezing operation (Fig. 12.13).

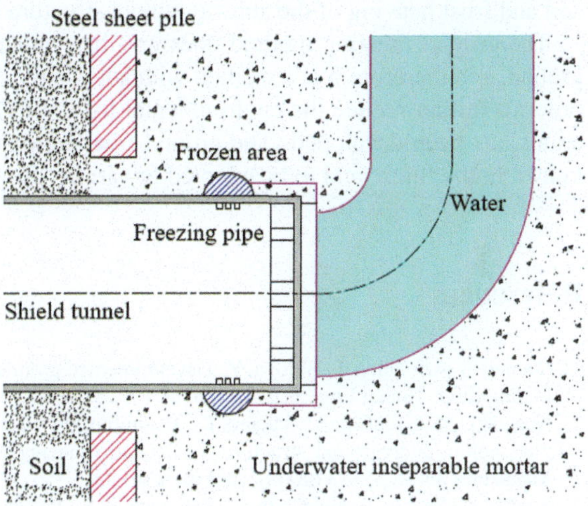

Fig. 12.12 Arrival protection work of shield machine [3]

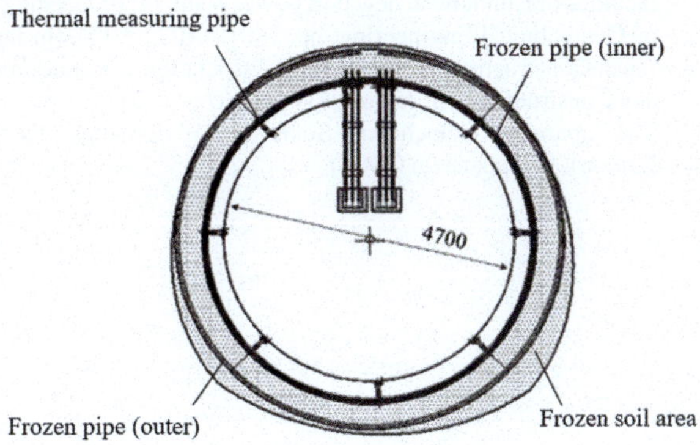

Fig. 12.13 Frozen soil area (After 20 days) [3]

12.5 Summary

The two projects introduced in this chapter are the case of the ground freezing method using the brine and the CO_2, respectively. In the protection work of the tunnel just beneath the subway tunnel structure by the brine system, existing steel sheet piles were removed safely considering the strength of frozen soil and frost heaving of the subway tunnel structure by the freezing. In the protection work of the tunnel beneath the subway tunnel structure by the brine system, existing steel sheet piles were removed safely, after calculation was done considering the strength of frozen soil and frost heaving of the subway tunnel structure by the freezing.

The ground freezing method does not inject materials such as cement into the ground, and therefore it is a method with a low environmental load. Furthermore, in order to further reduce the environmental load, a CO_2 system using natural refrigerants has been developed. The ground freezing method is expected to be widely known in the future and to be adopted in various projects.

References

1. Souma, S., Ari-izumi, T., Shioya, Y.: Development of ground freezing method by CO_2 gas-liquid mixing fluid. Tunnel Eng **26**, 2–10 (2016). (in Japanese)
2. Nagataki, M., Sadafuji, S., Sakano, K., Hasimoto, K., Hashimoto, K.: Removing of remaining piles below underground structure by Ground Freezing Method. Tunnel Undergr (Japanese Magazine) **39**(8), 35–33 (2008). (in Japanese)
3. Souma, S., Saitou, T., Hatakeda, D., Takayanagi, S., Watanabe, K., Abe, S., Shioya, Y.: Ground protection by CO_2 freezing for shield machine reaching—construction project of discharge facilities of ishiklaiwan new port power plant. In: Proceeding of 52nd Japan National Conference on Geotechnical Engineering, pp. 1401–1402 (2017), (in Japanese)
4. Japan Construction Mechanization Association.: Ground freezing method (Planning, Designing and Construction) (in Japanese) (1982)
5. The Japanese Geotechnical Society.: Soil freezing—its theory and practice- (1st Revised Edition), (in Japanese) (1994)

Chapter 13
Combined Ground Improvement Method for Cultural Heritage Preservation—A Case Study in Angkor Ruins, Cambodia

Yasushi Akazawa, Mitsuharu Fukuda, Yoshinori Iwasaki, and Juichi Nakazawa

13.1 Overview

In many cases, the ground improvement techniques are being used for the preservation of ruins made of soil or rocks. It is necessary to apply the techniques and the materials which are modified from the usual ground improvements according to the characteristics of ruins. Table 13.1 (Nakazawa et al. 2003) shows the list of ground improvement techniques and earthworks that are being applied for the ruin's preservation.

In this chapter, a case history of applying the ground improvement to the ruins preservation is reported. On Angkor project in Cambodia, the soil–lime was applied to construct the basement for the tower that was dismantled and rebuilt [1].

Y. Akazawa (✉)
Japan Cultural Heritage Consultancy Co., Ltd., Tokyo, Japan
e-mail: akazawa@b-hozon.co.jp

M. Fukuda
Taisei Geo-Tech Co., Ltd., Tokyo, Japan
e-mail: fuku564b@xug.biglobe.ne.jp

Y. Iwasaki
Geo-Research Institute, Osaka, Japan
e-mail: yoshi-iw@geor.or.jp

J. Nakazawa
Tokyo Engineering Consultants Co., Ltd., Osaka, Japan
e-mail: junakaza@gf7.so-net.ne.jp

H. Hazarika et al. (eds.), *Practices and Trends in Ground Improvement Techniques*, Developments in Geotechnical Engineering,
https://doi.org/10.1007/978-981-19-3322-6_13

Table 13.1 List of ground improvement techniques and earthworks

Objective targets	Ground improvement techniques	Other methods	
	Utilization of soil properties	Utilization of materials	Retaining wall Pipe-roofing Piling Tunneling Dewatering by well or bored holes, etc.
Embankment	Replacement Compaction	Soil–cement, Soil–lime Geosynthetics	
Slope and surface	Loading berm	Spraying Laying Filling and gluing Anchoring and Bolting	
Shallow foundation	Replacement	Light-weighting Mixing by stabilizers Anchoring and bolting	
Deep formation	Consolidation by dewatering	Grouting Grout Jetting Anchoring	

13.2 Background

Angkor ruins are situated at Siem Reap in Cambodia and are composed of many masonry structures of temples, moat, and large embankment, which were constructed from the 9th to the fifteenth centuries. These ruins were registered as Cultural Heritages Sites by UNESCO in 1992. At the same time, they were listed as endangered World Heritage Site. In 2004 they were removed from the list due to many conservation activities from around the world. However, they are still under dangerous states that may lead to collapse either by human activities or by natural forces. In this case study, the individual masonry towers of Prasat Suor Prat at Ankor Ruins were studied and investigated for the preservation work.

13.3 Project Outline and Objective Structure

Prasat Suor Prat in Angkor Thom is composed of 12 towers and terraces. The towers are made up of masonry structures and were constructed with laterite blocks. They are designed to look like three-storied structures. Each tower is constructed differently and has different structure problems, for example, opening of wall joints and leaning of towers, etc.

N1 tower was selected as the most dangerous tower among them for the restoration work (Fig. 13.1). The tower is 21 m in height and 9.2 m in width and 10.5 m in length at the base level. This tower inclined by about 6.6 degrees toward a nearby pond.

Fig. 13.1 N1 tower (Before reconstruction) [1]

13.4 Results of Investigations

To proceed with the design of N1 tower's restoration, case histories were collected on restoration of similar tower (N3 tower). Also, investigations on the existing states of N1 tower and the subsurface ground underneath were carried out. Based on these results, principles, guidelines, and methods for the N1 tower's restoration project were decided.

13.4.1 Lessons Learned from N3 Tower

N3 tower (one of Prasat Suor Prat) was restored by Ěcole francaise d'Extreme-Orient in 1950, and its superstructure was dismantled and rebuilt. Fifty years after its restoration, it inclined once again and structure materials for its apertures were damaged. It has been recognized that N3 tower is at risk of collapse. This leaning of N3 tower is supposed to have occurred during restoration work or just after it. It was thought to be caused by facts that only the superstructure was retrofitted, while its platform and its foundation ground were left untreated.

13.4.2 Survey of Existing State of N1 Tower

The change in N1 tower's inclination with time has been monitored by plumbing, however, no significant change was found. During the rainy season in October 1997, when there was wind velocity of more than 10 m/s for a few days, 0.5 mm/m of

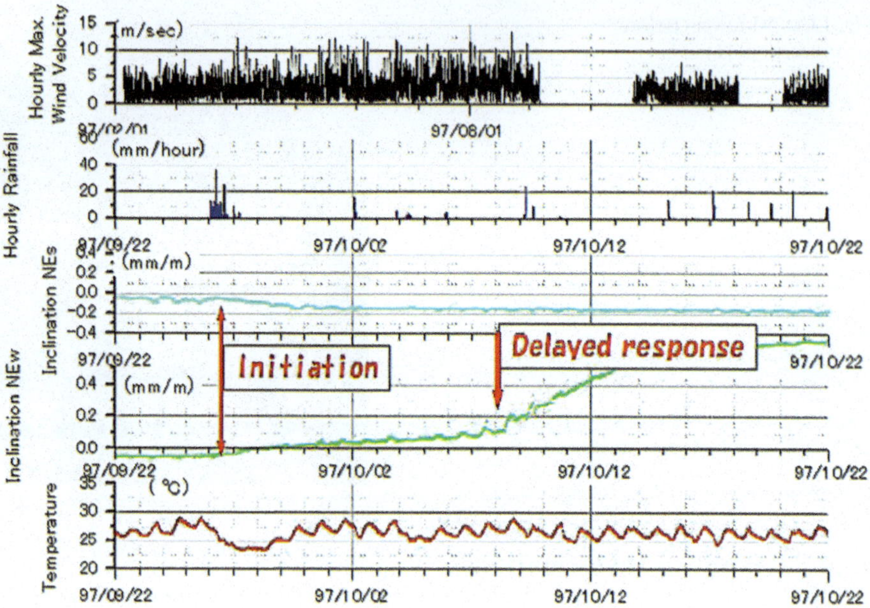

Fig. 13.2 The measuring results of the amount of inclinations, the wind velocity, precipitation, and temperature of N1 tower [1]

the lateral displacement was measured. Figure 13.2 shows this monitoring data with inclinations, wind velocity, rainfall, and temperature for the N1 tower. This fact shows that the tower's displacement was increased by the external force acting against the tower (in this case, gust wind during rainy season), and this displacement will be accumulated over time, which may cause potential collapse of the tower.

Based on these facts, it is recognized that the foundation platform and the ground underneath should be investigated in detail. First, a soil survey by hand auger was carried out, however, the results of it were insufficient to clarify the soil strata under the platform. It was concluded that the investigation of the tower should be done by dismantling them, including excavation of the foundation ground. Moreover, it was decided that these results will be used for the design of other ruins' restoration projects in the future.

Figures 13.3 and 13.4 show soil strata under the platform of N1 tower. The compacted sandy formation was found from −1.5 m from the original ground surface and to +3.5 m above the ground, in which the average thickness of one stratum was measured about 20 cm. Figure 13.5 shows the grain size distribution curves of them. Figure 13.6 shows N-value distribution (obtained from the standard penetration test) with depth. In the figure, white dots show the values in the dry season and the polygonal line shows the same in the rainy season.

The average in-situ dry density of the compacted strata was measured at 1.69 g/cm^3 at the center and 1.73 g/cm^3 directly under the laterite blocks. Because the maximum

Fig. 13.3 Soil strata under the platform of N1 tower [1]

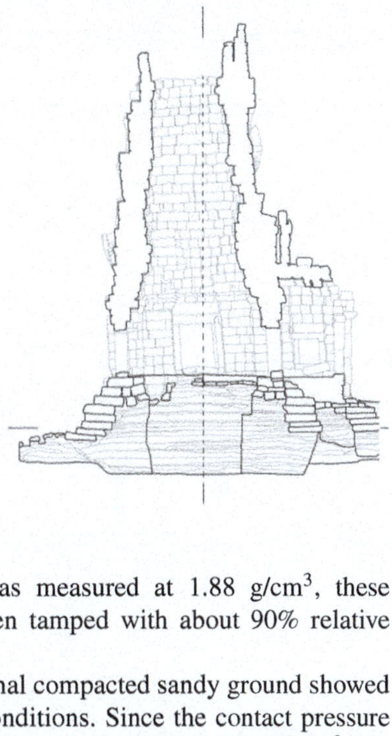

Fig. 13.4 North–South section of N1 tower (Before restoration) [1]

dry density by proctor's compaction test was measured at 1.88 g/cm^3, these compacting layers were assumed to have been tamped with about 90% relative compaction.

Plate bearing in-situ test of loading on original compacted sandy ground showed the yield load of 290–340kN/m^2 under wet conditions. Since the contact pressure in the ground beneath the foundation of the tower was estimated at 270kN/m^2, the safety factor against the ground bearing capacity in rainy season was calculated at SF = 1.1, which was considered to be small and unsafe. When the external force acts on the tower during the rainy season, the inclination, the subsidence, and differential settlement of them will increase.

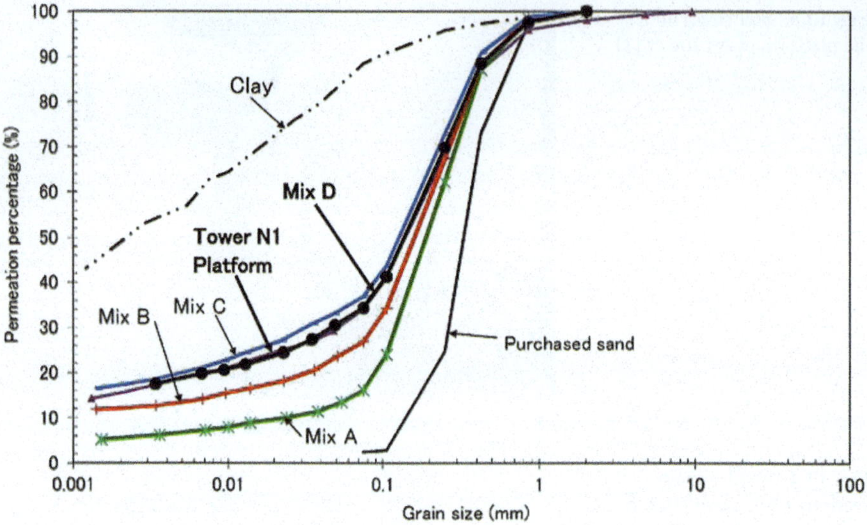

Fig. 13.5 Grain size distribution curve of each soil [1]

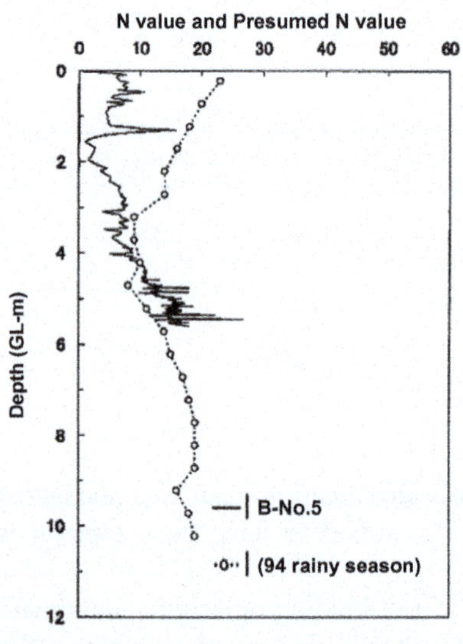

Fig. 13.6 N-value distribution [1]

13.5 Construction Method

The principle for this reconstruction is to adopt the original tamping method for excavated original soils. However, if the bearing layers are reconstituted by the original tamping method with the same soils, their bearing capacities will be same as the original states and bearing capacity reduction during rainy season will not able to be avoided. Therefore, it was concluded that the foundation ground should be improved with some stabilizing materials in the soil, if constructed with the original excavated soil.

In the case of in-situ surface soil stabilization, Portland cement or hydrated lime is used as a stabilizing agent. Hydrated lime was judged to be the best for this case because it is easy to buy near the site and also is easy to handle during working. Moreover, it is very durable.

13.6 Mixing Test of Slaked Lime Soil and Results

Table 13.2 shows the proportions of material soils to be prepared and slaked lime to be mixed for them. In Table 13.2, item "sand" means excavated sandy soil, which was used in the tower's construction. Each mixed soil was mixed with this "sand", clayey soil, and lateritic soil with weight ratio shown in Table 13.2. Slaked lime, which is a stabilizing material, is added to mixed soil 1 at the rates shown in Table 13.2. The mixing purpose of clayey soil was to increase the stabilization effect through chemical reaction of clay minerals and slaked lime. Also, in lateritic soil mixing, it was intended that the ferromagnetic ion and aluminum ion present in it would promote the chemical hardening reactions. Grain size accumulation curves of mixed soils are shown in Fig. 13.6. Results of the test were evaluated by comparing the unconfined compression strengths of the treated specimens.

(1) Unconfined compression strengths of the treated specimens were increased as the mixing rate of slaked lime was increased.
(2) The unconfined compression strengths and the deformation moduli of tested specimens increased with their curing times, and proportional relation between them was observed. Relation between the curing time and the unconfined compression strength of the mixed soil D is shown in Fig. 13.7.
(3) The changes in their dry densities by curing time were not recognized.

Based on above-mentioned results, the material soil decided to be used for the foundation platform and its base ground was a mixture of three different soils: the original sandy soil, clayey soil, and lateritic soil at weight ratios of 70, 15,

Table 13.2 Proportion of mixed soil

Items	Sand (excavated sandy soil)	Clayey soil	Lateritic soil	Slaked lime mixture rate (%)
Mixture A	100	0	0	5, 10, 15
Mixture B	85	15	0	5, 10, 15
Mixture C	70	30	0	5, 10, 15
Mixture D	70	15	15	5, 10, 15

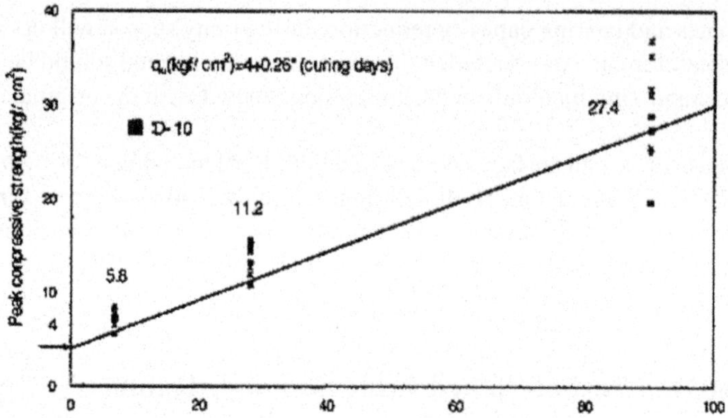

Fig. 13.7 Relation of unconfined compression strength and curing time [1]

and 15% (proportion of mixing soil D). Slaked lime, which was a stabilizing material, was added to mixed soil 1 at a rate of 0.1.

13.7 Construction Procedures

It was decided that mixed soil D proportion should be used for material soil and slaked lime should be used as a stabilizing material by adding to the mixed soil 1 at a rate of 0.1. The tamping method of construction was adopted. The construction procedures are as follows.

(1) The material soil treated with slaked lime should be spread at 15 cm thickness, and after that, it should be compacted to a thickness of 7.5 cm. By repeating this work, the foundation ground should be prepared to the designated level.

(2) Compacting work is decided by the field trial work to be composed of three steps. Primary compaction should be conducted using a wooden rod (dia. 3 cm) with manual tamping, and followed by secondary compaction using a flat steel plate (15 cm × 15 cm × 1 cm) in the same manner. Finally, surface compaction should be conducted by using tamping rammer.

(3) The designated foundation ground for the tower will be made of laminated structure with 7 cm thickness tamping layer. Since there is a smooth surface between each layer, existence of such layers may make the structure mechanically weak. In order to secure the bond of each layer, roughing work for each compacted surface should be done before tamping the next layer. The surface roughing work should be performed manually by using steel rakes.

When allowable bearing capacity is 1.2 MN/m^2, it is assumed that the compressive strength in place should be about 60% of the strength obtained in the laboratory test. This means that the target value should be 2 MN/m^2. In the case of maxed soil D, a curing time of more than 90 days would be required to achieve the target value of

2 MN/m^2. However since the reconstruction work for the superstructure needed to be executed without waiting for 90 days, some measures to reinforce the improved soil should be adopted until the onset of its strength. Therefore, it was decided that geotextile should be placed for temporally reinforcing the improved soils in order to achieve the required strength.

This design was applied on a section of the foundation ground shown in Fig. 13.8 three layers of geotextiles were inserted 3. Table 13.3 shows the specification of geotextile which was utilized. One of the effects of geotextile is to work as reinforced materials to increase the shear resistance of soil mass. It is expected that the use of geotextiles will increase the factor of safety of the sliding failure by about 20%.

On the other hand, when the curing time of treated soil is short, it was estimated that there will be insufficient bearing capacity of the foundation ground. Also, the foundation ground, which is composed of tamped layers and geotextiles, may not

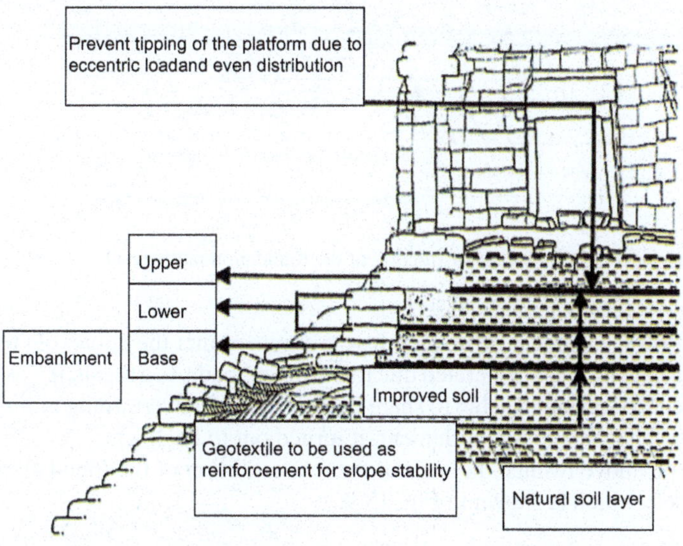

Fig. 13.8 Section of spreading Geotextile [1]

Table 13.3 The specification of geotextiles

Items	Values
Type	KJV-6000
Base texture	High-strength vinylon
Width	
	2000 mm
Weight	320 g/m^2
Tensile strength	59.0 kN/m
Reduction coefficient	0.6

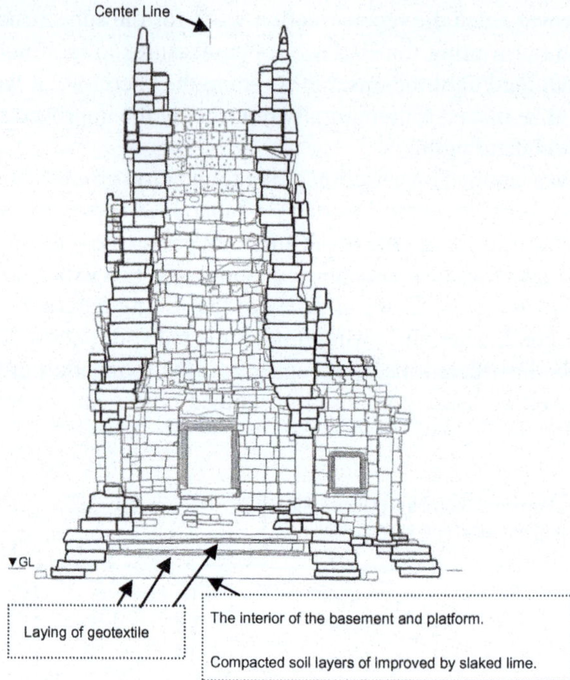

Fig. 13.9 Decided section for reconstruction of the foundation platform [1]

have uniform mechanical behaviors. It was assumed that the geotextile resists only the overturning moment generated due to the tower load. As a result, it was estimated that the geotextiles will be able to share 26% of the total overturning moment. When converting this value to the settlement, it turned out to be 40 cm.

Based on above results, the section for reconstitution of the foundation platform was decided, which is shown in Fig. 13.9.

13.8 Conclusion

This project was completed in May 2005. After that monitoring for these structures was done. After the reconstruction until now, there has been no trouble.

In recent years, consensus regarding how the cultural heritage should be conserved and saved from various collapsing factors is changing. If modern materials and modern methods are applied for restoration works of heritages, the historic value of them may be lost. [2] Therefore, it is demanded that materials and methods for the restoration works should be selected such that they will not destroy the historic values of heritages. From these perspectives, a concept called "authenticity and heritage" has emerged, especially regarding the design and execution of heritage geo-technology.

References

1. Akazawa,Y., Fukuda, M., Iwasaki, Y., Nakazawa., J.: Reconstitution of foundation platform of Prasat Suor Prat by compaction of original soil with slaked lime, Ankor Ruins Cambodia. In: Proceeding of the 18th International Conference on Soil Mechanics and Geotechnical Engineering, Paris (2013)
2. Nakazawa, J., Watanabe, Y., Akazawa, Y., Takamatu, M.: Ground improvement for preservation of historical ruins. In: Proceeding of the 54th Geotechnical Symposium of Japanese Geotechnical Society, pp. 543–550 (2009) (in Japanese)

Chapter 14
Epilogue

Hemanta Hazarika, Juichi Nakazawa, and Iwao Nakahara

The case histories described in this book include examples from both developing and developed countries. The purpose of selecting these case histories was to let the younger generation of engineers and researchers learn from these examples. Those knowledges can help them in selecting a particular ground improvement technique that best suits the varying geology, budget constraints, and available technologies, depending on the countries. Case histories need not be limited to selecting suitable methods, instrumentations of the sites, and evaluating the post improvement results. The challenges encountered and the lessons learned from the case history and passing the information to future generations are of vital importance.

On projects with budgetary constraints, adoption of innovative ground improvement techniques provides attractive and added benefits. Therefore, for designers confronted with varying site and loading conditions, adopting a flexible approach of combining two or more techniques on a single project could be of immense economic benefits.

On the other hand, a lot of scopes exist for researchers, who can contribute toward continuous development of new and innovative methods based on the latest technologies such as AI (Artificial intelligence), ICT (Information and Communication Technology), IoT (Internet of Things), DX (Digital Transformation), etc. In addition, we must look beyond the conventional materials for improving ground. The trend these

H. Hazarika (✉)
Kyushu University, Fukuoka, Japan
e-mail: hazarika@civil.kyushu-u.ac.jp

J. Nakazawa
Tokyo Engineering Consultants Co. Ltd., Osaka, Japan
e-mail: junakaza@gf7.so-net.ne.jp

I. Nakahara
Japan Foundation Engineering Co., Ltd., Tokyo, Japan
e-mail: iwao_nakahara@jafec.co.jp

H. Hazarika et al. (eds.), *Practices and Trends in Ground Improvement Techniques*,
Developments in Geotechnical Engineering,
https://doi.org/10.1007/978-981-19-3322-6_14

days is to utilize recycled materials: both natural and artificial. The artificial materials may include utilization of industrial by-products. New ideas and technologies also need to be developed toward realizing the SDGs set forth by the United Nations, and effectively engaging toward reducing the impact of climate change mentioned in the report of Intergovernmental Panel on Climate Change (IPCC).

The manufacturer's authorised representative in the EU is Springer Nature Customer Service Centre GmbH, Europaplatz 3, 69115 Heidelberg, Germany. If you have any concerns regarding our products, please contact ProductSafety@springernature.com

Printed and bound by CPI Group (UK) Ltd, Croydon, CR0 4YY
30/09/2026
02225308-0015